COGNITIVE SCIENCES
RESEARCH PROGRESS

COGNITIVE SCIENCES RESEARCH PROGRESS

MIAO-KUN SUN
EDITOR

Nova Science Publishers, Inc.
New York

Copyright © 2009 by Nova Science Publishers, Inc.

All rights reserved. No part of this book may be reproduced, stored in a retrieval system or transmitted in any form or by any means: electronic, electrostatic, magnetic, tape, mechanical photocopying, recording or otherwise without the written permission of the Publisher.

For permission to use material from this book please contact us:
Telephone 631-231-7269; Fax 631-231-8175
Web Site: http://www.novapublishers.com

NOTICE TO THE READER
The Publisher has taken reasonable care in the preparation of this book, but makes no expressed or implied warranty of any kind and assumes no responsibility for any errors or omissions. No liability is assumed for incidental or consequential damages in connection with or arising out of information contained in this book. The Publisher shall not be liable for any special, consequential, or exemplary damages resulting, in whole or in part, from the readers' use of, or reliance upon, this material.

Independent verification should be sought for any data, advice or recommendations contained in this book. In addition, no responsibility is assumed by the publisher for any injury and/or damage to persons or property arising from any methods, products, instructions, ideas or otherwise contained in this publication.

This publication is designed to provide accurate and authoritative information with regard to the subject matter covered herein. It is sold with the clear understanding that the Publisher is not engaged in rendering legal or any other professional services. If legal or any other expert assistance is required, the services of a competent person should be sought. FROM A DECLARATION OF PARTICIPANTS JOINTLY ADOPTED BY A COMMITTEE OF THE AMERICAN BAR ASSOCIATION AND A COMMITTEE OF PUBLISHERS.

LIBRARY OF CONGRESS CATALOGING-IN-PUBLICATION DATA

Available Upon Request

ISBN: 978-1-60456-392-4

Published by Nova Science Publishers, Inc. ✛ *New York*

CONTENTS

Preface — vii

Chapter 1   **Subcellular Trafficking of BACE1: Molecular Mechanisms in the Control of β-Amyloid Generation**   1
Tina Wahle and Jochen Walter

Chapter 2   **Brain Vesicular Monoamine Transporter and Apoptosis: Perspectives on Development and Neurodegeneration**   21
Léa Stankovski, Patricia Gaspar and Olivier Cases

Chapter 3   **Mild Cognitive Impairment is Too Late: The Case for Presymptomatic Detection and Treatment of Alzheimer's Disease**   33
Charles D. Smith

Chapter 4   **Semantically Mediated Integration of Cognition In *Homo Sapiens*: Evolution, Grammar, Uncertainty, and Cognitive Accuracy**   85
Charles E. Bailey

Index — 143

PREFACE

This book presents new research on cognitive science which is most simply defined as the scientific study either of mind or of intelligence. It is an interdisciplinary study drawing from relevant fields including psychology, philosophy, neuroscience, linguistics, anthropology, computer science, biology, and physics. There are several approaches to the study of cognitive science. These approaches may be classified broadly as symbolic, connectionist, and dynamic systems. Symbolic holds that cognition can be explained using operations on symbols, by means of explicit computational theories and models of mental (but not brain) processes analogous to the workings of a digital computer. Connectionist (subsymbolic) holds that cognition can only be modeled and explained by using artificial neural networks on the level of physical brain properties. Hybrid systems hold that cognition is best modeled using both connectionist and symbolic models, and possibly other computational techniques. Dynamic Systems hold that cognition can be explained by means of a continuous dynamical system in which all the elements are interrelated, like the Watt Governor.

The essential questions of cognitive science seem to be: What is intelligence? and How is it possible to model it computationally?

Chapter 1 - Alzheimer's disease (AD) is the most common neurodegenerative disorder, characterized by synaptic loss, and the combined occurrence of neurofibrillary tangles and amyloid plaques in the brain. While neurofibrillary tangles are intraneuronal aggregates of the microtubuli-associated protein tau, amyloid plaques are extracellular deposits containing the amyloid peptide (A). A derives from the amyloid precursor protein (APP) by sequential proteolytic cleavage by and γ-secretase. Secretase has been identified as the aspartic protease BACE1 that cleaves APP at the N-terminus of the A domain generating a membrane-bound C-terminal fragment (CTF). CTF serves as a substrate for γ-

secretase that mediates cleavage within the transmembrane domain resulting in the release of A. Because BACE1 initiates A generation, it represents a potential target molecule to decrease A production in therapeutic strategies for AD.

BACE1 is a type I membrane protein that is transported in vesicular compartments of the secretory and endocytic pathway. Consistent with the acidic pH optimum of BACE1, the secretory processing of APP occurs predominantly in compartments of the endocytic pathway. The small cytoplasmic domain of BACE1 is critical in the regulation of its subcellular transport and undergoes post-translational modifications by palmitoylation and phosphorylation. Moreover, the cytoplasmic domain interacts with proteins of the GGA family and the copper chaperone for superoxide dismutase-1. Here, the authors review the molecular mechanisms involved in the regulation of the subcellular trafficking and activity of BACE1, and discuss the potential involvement in the pathogenesis of AD.

Chapter 2 - The neuronal isoform of vesicular monoamine transporter, VMAT2, is responsible for packaging dopamine, norepinephrine, and serotonin into synaptic vesicles and thereby plays an essential role in monoamine neurotransmission. This sequestering action is important for normal synaptic release of monoamines but it may also act to keep intracellular levels of the monoamine transmitters below potentially toxic levels. During embryonic and postnatal development, VMAT2 is expressed in many non-aminergic neurons in the brain, long before synapses are formed, suggesting that it could have non-synaptic roles. Here, the authors review recent evidences indicating a role for VMAT2 in the control of developmental cell death in the cerebral cortex and in models of Parkinson related disorders. Abnormalities in VMAT2 functions have been suggested to play a key role in the etiology of a number of disorders, including Parkinson's disease and addiction.

Chapter 3 - The thesis of this review is that the earliest cognitive symptoms in dementia represent the exhaustion of compensatory mechanisms in the brain which counteract underlying Alzheimer's disease (AD), vascular, Lewy body, and other neuropathology. These pathologies may accumulate gradually over years and perhaps enter a phase of acceleration as compensatory resources are outstripped. Evidence for this model of a prolonged presymptomatic period in AD has come from recent imaging, neuropathologic, and basic science studies. These studies and the potential consequences for diagnosis and treatment are presented. The conclusion is that using onset symptoms as the signal to begin disease-modifying treatment for AD is too late; this treatment must begin earlier, before symptoms begin, to preserve brain function. Therefore presymptomatic detection is a critical research goal in AD.

Chapter 4 - This article reviews research on human brain, cognition, language, behavior, and evolution to posit the value of operating with a stable reference point based on cognitive accuracy and a rational bias. Drawing on rational emotive, cognitive behavioral and cognitive neuroscience on the one hand and a general brain model of frontal lobe executive function and working memory on the other, along with proposed language mediation of cognitive processes, this review yields potential implications for maximizing brain functioning of *Homo sapiens*. Cognitive thought processes depend on the operations and interactions of specific brain structures and networks, functioning more effectively under conditions of cognitive accuracy (including accurate information, thought process accuracy, and event-level accuracy). However, typical cognitive processes appear to promote the adoption and use of subjective cultural beliefs, mediated by language and grammatical habits mostly learned during early development. In turn, these grammatical habits tend to bias humans toward cognitive inaccuracies. On the other hand, a process that applies informed frontal lobe executive functioning to the mediation of cognition, emotion, and behavior may help to minimize the negative effects of indiscriminately applied cultural belief systems, provide a naturalistic framework for future research and ultimately enhance cognitive accuracy as a reference point for evaluating humans while offering improved relative environmental homeostasis.

In: Cognitive Sciences Research Progress
Editor: Miao-Kun Sun

ISBN: 978-1-60456-392-4
© 2008 Nova Science Publishers, Inc.

Chapter 1

SUBCELLULAR TRAFFICKING OF BACE1: MOLECULAR MECHANISMS IN THE CONTROL OF β-AMYLOID GENERATION

Tina Wahle and Jochen Walter[*]

Department of Neurology, University of Bonn, Sigmund-Freud-Str. 25, 53127 Bonn, Germany

ABSTRACT

Alzheimer's disease (AD) is the most common neurodegenerative disorder, characterized by synaptic loss, and the combined occurrence of neurofibrillary tangles and amyloid plaques in the brain. While neurofibrillary tangles are intraneuronal aggregates of the microtubuli-associated protein tau, amyloid plaques are extracellular deposits containing the amyloid peptide (A). A derives from the amyloid precursor protein (APP) by sequential proteolytic cleavage by and γ-secretase. Secretase has been identified as the aspartic protease BACE1 that cleaves APP at the N-terminus of the A domain generating a membrane-bound C-terminal fragment (CTF). CTF serves as a substrate for γ-secretase that mediates cleavage within the transmembrane domain resulting in the release of A. Because BACE1 initiates A generation, it represents a potential target molecule to decrease A production in therapeutic strategies for AD.

---

[*] Correspondence and requests for reprints: Jochen Walter, Ph.D., Department of Neurology, University of Bonn , Sigmund-Freud-Str. 25, 53127 Bonn . Tel: + 49 228 19782; Fax: +49 228 14387; Email: Jochen.Walter@ukb.uni-bonn.de

BACE1 is a type I membrane protein that is transported in vesicular compartments of the secretory and endocytic pathway. Consistent with the acidic pH optimum of BACE1, the secretory processing of APP occurs predominantly in compartments of the endocytic pathway. The small cytoplasmic domain of BACE1 is critical in the regulation of its subcellular transport and undergoes post-translational modifications by palmitoylation and phosphorylation. Moreover, the cytoplasmic domain interacts with proteins of the GGA family and the copper chaperone for superoxide dismutase-1. Here, we review the molecular mechanisms involved in the regulation of the subcellular trafficking and activity of BACE1, and discuss the potential involvement in the pathogenesis of AD.

INTRODUCTION

Alzheimer's disease (AD) is neuropathologically characterized by the occurrence of neurofibrillary tangles and β-amyloid plaques in the brain [Selkoe, 2001]. Major components of plaques are amyloid β-peptides (Aβ) that derive from the β-amyloid precursor protein (APP) by endoproteolytic processing involving sequential cleavages by β- and γ-secretase [Selkoe, 2001; Annaert et al., 2002; Walter et al., 2001]. The initial cleavage of APP by β-secretase generates a membrane-bound C-terminal fragment that contains the Aβ domain (CTFβ). This fragment represents a substrate for γ-secretase, which mediates the apparently intramembraneous cleavage of APP resulting in the liberation of Aβ. In an alternative pathway, APP can be cleaved by α-secretase within the Aβ domain thereby precluding the generation of Aβ [Selkoe, 2001; Annaert et al., 2002; Walter et al., 2001].

The proteolytic processing of APP by α-, β- and γ-secretases occurs predominantly in post-Golgi secretory and endocytic compartments, and at the cell surface [Annaert et al., 2002; Walter et al., 2001; Steiner et al., 2000]. The generation of Aβ is not only determined by the relative expression levels of APP and secretases, but also by their subcellular localization. APP as well as the secretases are integral membrane proteins and are transported in the secretory pathway from the endoplasmic reticulum (ER) to the cell surface, from where they can be internalized into endosomal/lysosomal compartments [Koo et al., 1996; Yamazaki et al., 1996, Walter et al., 1996; Walter et al., 2001; Annaert et al., 1999]. Thus, enzyme–substrate interactions could potentially occur in distinct subcellular compartments. By cell biological and biochemical experiments it was demonstrated that β-secretase cleaves wild-type APP predominantly in endosomal/lysosomal compartments after its re-internalization from the cell surface [Perez et al., 1996; Lo et al., 1994; Haass et al., 1995]. Notably, the Swedish double mutation in APP that is located directly at the cleavage site for β-secretase and causes early onset AD, significantly increases the production of Aβ by enhancing the cleavage efficiency by BACE1 in the secretory pathway [Citron et al., 1992; Haass et al., 1995; Thinakaran et al., 1996].

PHYSIOLOGICAL AND PATHOPHYSIOLOGICAL FUNCTIONS OF THE β-SECRETASE BACE1

β-secretase was identified as the aspartic protease BACE1 (beta site APP cleaving enzyme) also called Asp-2 (aspartic protease-2) or Memapsin-2 (membrane-associated aspartic protease-2) [Vassar et al., 1999; Sinha et al., 1999; Hussain et al., 1999; Yan et al., 1999]. BACE1 together with its close homolog BACE2 form a subfamily of transmembrane aspartic proteases related to the pepsin family [Bennet et al., 2000]. While the BACE1 gene is localized on chromosome 11, that of BACE2 is localized on chromosome 21 within the critical Down's syndrome region [Saunders et al., 1999]. However, whether BACE2 contributes to the pathogenesis of Down's syndrome or AD remains unclear.

Both enzymes contain the two characteristic D(T/S)G(T/S) motifs of aspartyl proteases, which form their catalytic site [Vassar et al., 2001; Walter et al., 2001]. In contrast to all other aspartyl proteases of the pepsin family, BACE1 and BACE2 are type I transmembrane proteins with a large lumenal domain, a single transmembrane domain and a small cytoplasmic tail (Fig. 1). Although, both proteases share homology of ~ 75%, strong evidence supports a predominant role of BACE1 in the generation of Aβ.

(1) The expression of BACE1 mRNA in neurons which predominantly produce Aβ in the brain, is much higher than that of BACE2 [Sinha et al., 1999; Vassar et al., 1999; Bennett et al., 2000].
(2) Aβ production is almost completely inhibited in mice lacking BACE1 [Luo et al., 2001; Cai et al., 2001].
(3) Antisense inhibition of BACE1, but not of BACE2, results in decreased β-secretase cleavage of APP [Vassar et al., 1999; Yan et al.; 1999].
(4) BACE1 cleaves at the β-secretase site(s) of APP much more efficiently than BACE2 [Sinha et al., 1999; Vassar et al., 1999; Yan et al.; 1999; Farzan et al., 2000].

Moreover, elevated BACE1 protein levels and enzymatic activities have been reported in the brains of AD patients [Holsinger et al., 2002; Yang et al., 2003]. However, BACE2 could also cleave APP at the β-secretase site [Farzan et al., 2000; Fluhrer et al., 2002]. Also, glial cells of BACE1 KO mice, but not of BACE1/BACE2 double KO mice could still produce Aβ, indicating that BACE2 could contribute to Aβ generation [Dominguez et al., 2005].

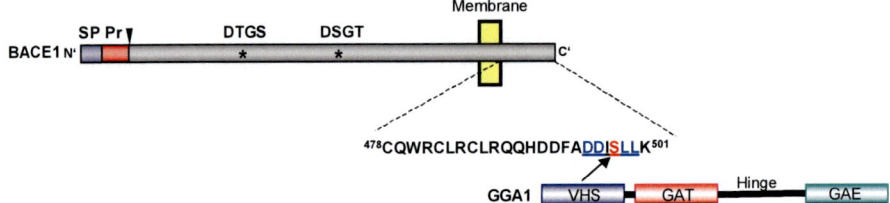

Figure 1: Schematic of BACE1. Models of BACE1 and GGA1. The aspartyl protease active site motifs (DTGS, DSGT) in the lumenal/extracellular domain of BACE1 is indicated. The amino acid sequence of the cytoplasmic domain of BACE1 is given in single letter code. The interaction motif of the BACE1 cytoplasmatic domain for GGA1 is in blue, the phosphorylation site of BACE1 (S 498) is in red. The functional domains of GGA1 are indicated. GGA1 interacts with the cytoplasmic domain of BACE1 via the VHS domain (arrow; see text for details).

Besides APP, additional candidate BACE1 substrates were identified, including the sialyltransferase ST6Gal I [Kitazume et al, 2001], the adhesion protein P-selectin glycoprotein ligand-1(PSGL-1) [Lichtenthaler et al, 2003], β-subunits of voltage gated sodium channels (Wong et al., 2005), APP-like proteins (APLPs) [Lie et al., 2003], low density lipoprotein-receptor-related protein (LRP) [von Arnim et al, 2005] and the interleukin-1 receptor II (IL-1R2) [Kuhn et al., 2007]. Notably, BACE1 can also cleave Aβ at position 34, leading to decreased levels of Aβ40 in conditioned media of cultured cells [Fluhrer et al., 2003; Shi et al., 2003]. In line with these findings, transgenic mice with high overexpression of BACE1 showed reduced plaque load in the brain, indicating an Aβ degrading activity of BACE1 [Lee et al., 2005]. However, the biological relevance of these findings remain unclear.

Initial studies with BACE1 KO mice did not reveal an overt phenotype (Luo et al., 2001; Roberds et al, 2001). However, other BACE1 KO mouse models showed increased lethality after birth and subtle electrophysiological and behavioral alterations [Domiguez et al., 2005; Laird et al., 2005]. Recently, it has been shown that BACE1 is selectively expressed at high levels in postnatal days 0-17 correlating with the myelination period of peripheral neurons [Willem et al.,

2006]. The analysis of BACE1 KO mice revealed hypomyelination and axonal bundling abnormalities, associated with accumulation of type III neuregulin-1. Alterations in myelin associated proteins and accumulation of neuregulin-1 were also observed in cortical and hippocampal neurons, indicating that BACE1 is also involved in myelination of nerve cells in the central nervous system [Hu et al., 2006]. In line with this, the activity of BACE1 has been shown to be negatively regulated by NOGO, that also associates with myelin (He et al., 2004). It has also been shown that, despite decreased Aβ levels, high overexpression of BACE1 induced overt axonal degeneration associated with thinning of the myelin sheets of CNS neurons [Rockenstein et al., 2005]. Together, these data indicate that BACE1 could play important roles in neuronal function, independent of APP processing and Aβ generation.

THE SUBCELLULAR TRANSPORT OF BACE1

In peripheral cells, BACE1 is mainly localized in endosomal, lysosomal and TGN compartments [Vassar et al., 1999; Sinha et al., 1999; Hussain et al., 1999; Yan et al., 1999]. The characterization of the subcellular transport of BACE1 revealed that the secretase is transported in the secretory pathway from the ER to the Golgi compartment. After translocation into the ER, the lumenal domain of BACE1 is N-glycosylated at four asparagine residues and three intramolecular disulfide bonds are formed [Haniu et al., 2000]. Mutagenesis experiments revealed a critical role of disulfide bond formation for the catalytic activity of BACE1 [Haniu et al., 2000]. Shortly after the release of BACE1 from the ER and during further transport through the Golgi, its N-terminal prodomain is endoproteolytically removed by furin and/or a furin-like protease to generate the mature enzyme [Capell et al., 2000; Bennet et al., 2001; Benjannett et al., 2001; Creemers et al., 2000]. The prodomain does not strongly inhibit enzyme activity, as one might expect for a typical zymogen, but rather plays a role in folding and forward transport of BACE1 through the secretory pathway [Benjannett et al., 2001; Shi et al., 2001]. In the Golgi compartment, BACE1 also undergoes complex modification of the N-linked glycostructures [Capell et al., 2000]. After complete maturation, the enzyme is further transported from the TGN to the cell surface from where it can be re-internalized into early endosomal compartments [Walter et al., 2001; Huse et al., 2000]. From early endosomes, BACE1 can either recycle to the cell surface, be transported to later endosomal compartments or retrieved to the TGN [Walter et al., 2001; Huse et al., 2000]. In addition, BACE1 could be directed into the lysosomal pathway as well as in the ubiquitin-proteasom system for degradation [Quing et al., 2004; Koh et al., 2005].

In polarized Madin-Darby canine kidney (MDCK) cells, the majority of BACE1 is sorted to the apical surface while APP undergoes basolateral sorting. However, a significant fraction of BACE1 is sorted basolaterally where it competes with basolaterally sorted α-secretase for cleavage of APP. Thus, Aβ production in polarized cells is determined by the differential targeting of BACE1 and its substrate APP [Capell et al., 2002].

Studies with rat hippocampal neurons revealed that BACE1 is targeted to axons and somato-dendritic compartments as well as presynaptic terminals [Capell et al., 2000]. It was proposed that BACE1 is axonally transported together with its substrate APP and γ-secretase in the same vesicles which would allow Aβ generation in these transport compartments [Kamal et al., 2000]. However, life cell imaging of primary neurons showed that APP and BACE1 carrier vesicles have distinct characteristics in the axonal transport, indicating that the two proteins could be transported in different types of vesicles [Goldsbury et al., 2006]. These data are corroborated by studies with sciatic nerves of mice showing that PS1 and BACE1 are transported separately from APP [Lazarov et al., 2005]. It will therefore be interesting to further characterize the subcellular sites of BACE1 dependent processing of APP and Aβ generation in neuronal cells in more detail.

THE ROLE OF THE CYTOPLASMATIC DOMAIN OF BACE1 IN ITS SUBCELLULAR TRAFFICKING

The cytoplasmic domain of BACE1 contains several amino acid based trafficking signals (Fig. 1). A dileucine motif has been shown to regulate endocytosis and recycling of BACE1 to the plasma membrane, and deletion of this motif increased the levels of BACE1 at the cell surface [Huse et al., 2000]. In addition, a phosphorylation site for casein kinase 1 at Ser498 has been identified that regulates the retrograde trafficking of BACE1 between endosomal compartments and the TGN (Fig. 1; [Walter et al., 2001; Pastorino et al., 2002]). While phosphorylated BACE1 is efficiently retrieved from endosomal compartments to the TGN, a mutant that can not be phosphorylated accumulated in endosomal compartments.

BACE1 can also undergo palmitoylation at three cystein residues within the cytosolic tail. Although the exact role of BACE1 palmitoylation remains to be elucidated, this modification could affect its transport and/or distribution in detergent resistant membrane microdomains (DRMs) [Benjannett et al., 2001]. DRMs might function in the trafficking of proteins in the secretory and endocytic pathways in epithelial cells and neurons, and participate in a number of important biologic functions [Simons et al., 2000]. Interestingly, BACE1 was shown to associate with DRMs where it could preferentially process APP [Ehehalt et al, 2003]. In addition, artificial GPI-anchoring of BACE1 increased the DRM association and increased β-secretory processing of APP [Cordy et al., 2003].

The cytoplasmic domain of BACE1 can also interact with the copper chaperone for copper–zinc superoxide dismutase 1 (CCS) which is involved in post-transcriptional SOD1 activation by delivering Cu^+ to the enzyme [Angeletti et al., 2005]. The CCS and SOD1 heterodimer becomes linked via a disulfide bond, which is thought to be a prerequisite for SOD1 activation by Cu^+ and Zn^{2+} [Wong et al., 2000]. SOD1 acts as an anti-oxidant enzyme by lowering the steady-state concentration of superoxide, and mutations are associated with familial amyotrophic lateral sclerosis [Valentine et al., 2003; Angeletti et al., 2005]. It has been suggested that BACE1 competes with SOD-1 for binding to CCS, but a physiological relevance of this interactions remains to be determined. The recently identified RNA aptamer that selectively blocks the binding of CCS by competition with the CCS binding site of the cytoplasmic domain of BACE1 might help to clarify the role of this interaction in cellular models [Rentmeister et al., 2006].

In addition to the interaction with other proteins, BACE1 can also undergo homodimerization in cultured cells and human brain tissue [Schmechel et al., 2004; Westmeyer et al., 2004]. The dimerization is mediated by the transmembrane and the cytoplasmic domains and increases BACE1 activity in cleavage of APP [Schmechel et al., 2004; Westmeyer et al., 2004].

The cytoplasmic domain of BACE1 also contains a characteristic DXXLL motif for the interaction with Golgi-localized gamma ear containing (ADP) ribosylation factor binding (GGA) proteins (Fig. 1), that mediate sorting of specific cargo proteins, like the mannose-6-phosphate receptors, from the TGN to endosomal/lysosomal compartments [Bonifacino et al., 2004; Robinson et al, 2004]. Several groups demonstrated that GGAs also directly interact with BACE1 [He et al., 2002; von Arnim et al., 2004; He et al., 2005; Wahle et al., 2005]. Consistent with the lack of a DXXLL motif, BACE2 did not bind to GGA proteins [He et al., 2002; Wahle et al., 2005]. Notably, the interaction of GGA proteins with BACE1 is strongly increased by phosphorylation of serine residue 498 within the DISLL motif of the BACE1 cytoplasmic domain. Thus, GGAs could regulate the phosphorylation-state dependent subcellular transport of

BACE1. By using dominant-negative variants and RNAi mediated downregulation of GGAs, an involvement of these adaptor proteins in the endodcytic trafficking of BACE1 has been established [He et al., 2005; Wahle et al., 2005]. Specifically, GAA1 appears to mediate the phosphorylation dependent retrograde transport of BACE1 from endosomal compartments to the TGN, while non-phosphorylated BACE1 could enter a direct recycling route to the cell surface [Wahle et al., 2005].

HOW COULD GGA 1 DEPENDENT TRAFFICKING OF BACE1 INFLUENCE Aβ GENERATION ?

Several studies indicated that the secretion of APPs and Aβ was not affected by GGA proteins in cells overexpressing BACE1 [He et al., 2005; von Arnim et al., 2004, Wahle et al., 2005]. This might be partly attributable to aberrant cleavage of APP by overexpressed BACE1. Indeed, it has been shown that overexpression of BACE1 strongly increases β-secretory cleavage of APP already in early secretory compartments [Capell et al., 2000; Liu et al., 2002; Lee et al., 2005], whereas under physiological expression levels, β-secretase cleavage of wild-type APP has been localized predominantly to endosomal/lysosomal compartments [Koo et al., 1994; Haass et al., 1995; Thinakaran et al., 1996]. High overexpression of BACE1 also decreases Aβ levels in cell culture and transgenic mice by additional cleavages within Aβ [Fluhrer et al., 2002, 2003; Liu et al., 2002; Lee et al., 2005]. Thus, overexpression of BACE1 and aberrant cleavage of APP might have masked GGA-dependent effects on Aβ generation in post-Golgi secretory and endocytic compartments.

Indeed, functional analyses of cultured cells with endogenous BACE1 expression demonstrated that GGA1 is implicated in the proteolytic processing of APP. Over-expression of GGA1 reduced cleavage of APP by BACE1 as indicated by a decrease in CTFβ generation. Notably, GGA1 expression also reduced the secretion of Aβ while depletion of GGA1 by siRNA increased the generation of Aβ (Fig. 2; [Wahle et al. 2006; von Arnim et al., 2006]). Since GGA1 did not directly bind to APP, these data indicate that changes in the subcellular trafficking of BACE1 or other GGA1-dependent proteins contribute to changes in APP processing and Aβ generation [Wahle et al., 2006]. Interestingly, expression of GGA1 also decreased the γ-secretase cleavage of APP, suggesting that GGAs may affect proteolytic processing of APP at distinct steps [von Arnim et al., 2006].

GGAs were initially identified to mediate forward transport of cargo proteins from the TGN to endosomal/lysosomal compartments [Bonifacino et al., 2004]. Whether BACE1 is also transported on this route remains to be determined. However, GGA proteins are also involved in the retrograde transport of BACE1 from endosomes to the TGN [Wahle et al., 2005]. Since BACE1 can cleave APP in both compartments to initiate Aβ generation, the GGA dependent transport between the TGN and endosomes could likely regulate the amyloidogenic processing of APP by affecting the relative contribution of α- and β-secretory processing of APP (Fig. 2).

GGA proteins could also affect the proteolytic processing of APP independent of a direct interaction with BACE1. It was demonstrated that LR11/SorLA interacts with APP and affects its subcellular localization and proteolytic processing [Andersen et al., 2006; Offe et al., 2006; Spoelgen et al., 2006]. Because LR11/SorLA also represents cargo for GGA proteins [Jacobsen et al., 2002], it is plausible that trafficking and metabolism of APP could be affected when LR11/SorLA is complexed with GGA proteins. In addition, GGA proteins are adaptor proteins that, beside cargo protein binding, also interact with the small GTPase ARF, clathrin, and additional proteins involved in vesicle assembly and targeting [Bonifacino, 2004]. Therefore, expression of GGA1 variants might also impair vesicular transport by interfering with ARF- or clathrin-dependent mechanisms that could also alter APP processing. Moreover, GGAs could affect APP metabolism indirectly by regulating the transport of other proteins. Because GGAs mediate forward trafficking of the M6PR from the TGN to endosomal compartments [Puertollano et al., 2001], and expression of this receptor has also been shown to affect Aβ generation [Mathews et al., 2002], the GGA dependent changes in M6PR trafficking could also contribute to the observed changes in APP processing.

CONCLUSIONS

In the human brain, GGA1 is preferentially expressed in neurons. Notably, total expression of GGA1 is significantly decreased in AD brain [Wahle et al., 2006]. The reasons for reduced GGA1 expression in these individuals remain to be determined, but the described findings strongly suggest that low levels of the adaptor protein may contribute to increased Aβ production which could accelerate formation of senile plaques in sporadic AD. Thus, altered GGA1 expression and activity may be a risk factor for sporadic AD.

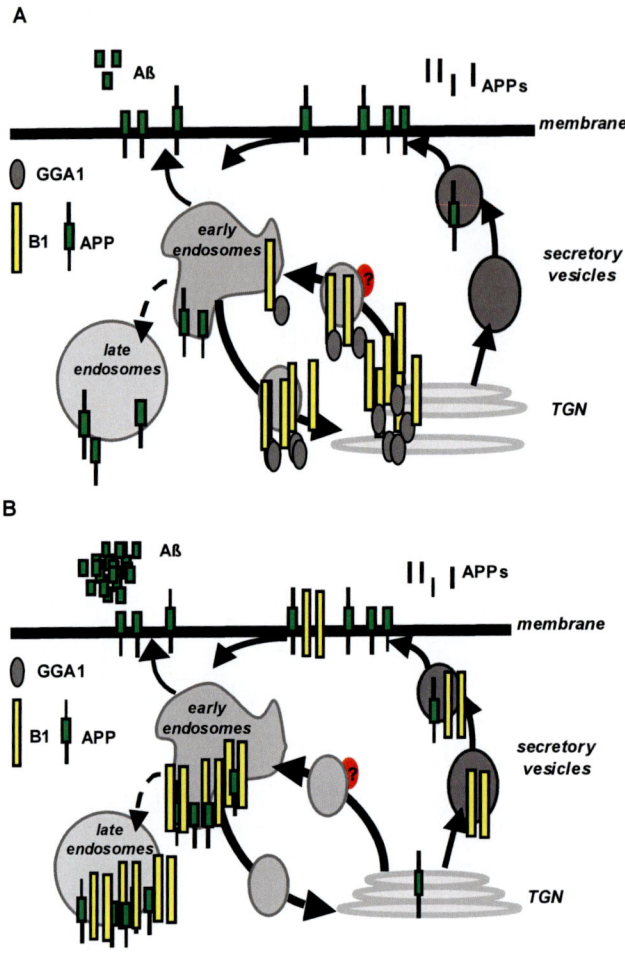

Figure 2: Model for the role of GGA1 in APP processing. Hypothetic model for the role of GGA1 in the subcellular trafficking of BACE1 and processing of APP in cells with higher (A) or lower (B) expression of GGA1. BACE1 and APP share similar transport routes within the secretory pathway to the cell surface (1), and both proteins could be endocytosed from the cell surface into endosomal compartments (2). GGA1 is involved in the retrograde transport from endosomal compartments to the TGN (3), and could thereby reduce the amount of BACE1 in these compartments. Consistent with this model, overexpression of GGA1 in cultured cells reduced processing of APP to Aβ. In addition, GGA1 might also facilitate the direct transport of BACE1 from the TGN to endosomal compartments. Decreased expression of GGA1 by RNAi led to elevated secretion of Aβ probably by increasing the amount of BACE1 in endosomal compartments. Additional GGA-dependent mechanisms, e.g., interaction with LR11/SorLA or M6P receptor (?), could also affect trafficking the proteolytic processing of APP (see text for details).

Recently, LR11/SorLA that is also involved in the subcellular transport of APP turned out to be a risk factor for sporadic AD [Andersen et al., 2005; Rogaeva et al., 2007]. Overexpression of LR11/SorLA in neurons caused decreased generation of Aβ, whereas ablation of LR11/SorLA expression in knockout mice resulted in increased levels of Aβ in the brain, similar to the situation in AD patients [Andersen et al., 2005]. These data suggest that inherited or acquired changes in LR11/SorLA expression or function are involved in the pathogenesis of AD [Rogaeva et al., 2007]. Altered trafficking of APP or its proteases has already been shown to be an important risk factor for sporadic AD and therefore may represent an effective therapeutic approach. It will therefore be interesting to further investigate the potential of GGA proteins and other factors involved in the subcellular trafficking of APP and secretases as targets for therapeutic intervention in AD.

ACKNOWLEDGEMENTS

We thank Kai Prager and Andrea Rentmeister for the helpful discussion on the manuscript. This work has been supported by grants of the Deutsche Forschungsgemeinschaft to J.W.

REFERENCES

Andersen, O. M., Reiche, J., Schmidt, V., Gotthardt, M., Spoelgen, R., Behlke, J., von Arnim, C. A., Breiderhoff, T., Jansen, P., Wu, X. et al. (2005). Neuronal sorting protein-related receptor sorLA/LR11 regulates processing of the amyloid precursor protein. *Proc Natl Acad Sci U S A* 102, 13461-6.

Angeletti, B., Waldron, K. J., Freeman, K. B., Bawagan, H., Hussain, I., Miller, C. C., Lau, K. F., Tennant, M. E., Dennison, C., Robinson, N. J. et al. (2005). BACE1 cytoplasmic domain interacts with the copper chaperone for superoxide dismutase-1 and binds copper. *J Biol Chem* 280, 17930-7.

Annaert, W. and De Strooper, B. (2002). A cell biological perspective on Alzheimer's disease. *Annu Rev Cell Dev Biol* 18, 25-51.

Annaert, W. G., Levesque, L., Craessaerts, K., Dierinck, I., Snellings, G., Westaway, D., George-Hyslop, P. S., Cordell, B., Fraser, P. and De Strooper, B. (1999). Presenilin 1 controls gamma-secretase processing of amyloid

precursor protein in pre-golgi compartments of hippocampal neurons. *J Cell Biol* 147, 277-94.
Benjannet, S., Elagoz, A., Wickham, L., Mamarbachi, M., Munzer, J. S., Basak, A., Lazure, C., Cromlish, J. A., Sisodia, S., Checler, F. et al. (2001). Post-translational processing of beta-secretase (beta-amyloid-converting enzyme) and its ectodomain shedding. The pro- and transmembrane/cytosolic domains affect its cellular activity and amyloid-beta production. *J Biol Chem* 276, 10879-87.
Bennett, B. D., Babu-Khan, S., Loeloff, R., Louis, J. C., Curran, E., Citron, M. and Vassar, R. (2000). Expression analysis of BACE2 in brain and peripheral tissues. *J Biol Chem* 275, 20647-51.
Bennett, B. D., Denis, P., Haniu, M., Teplow, D. B., Kahn, S., Louis, J. C., Citron, M. and Vassar, R. (2000). A furin-like convertase mediates propeptide cleavage of BACE, the Alzheimer's beta -secretase. *J Biol Chem* 275, 37712-7.
Bonifacino, J. S. (2004). The GGA proteins: adaptors on the move. *Nat Rev Mol Cell Biol* 5, 23-32.
Cai, H., Wang, Y., McCarthy, D., Wen, H., Borchelt, D. R., Price, D. L. and Wong, P. C. (2001). BACE1 is the major beta-secretase for generation of Abeta peptides by neurons. *Nat Neurosci* 4, 233-4.
Capell, A., Meyn, L., Fluhrer, R., Teplow, D. B., Walter, J. and Haass, C. (2002). Apical sorting of beta-secretase limits amyloid beta-peptide production. *J Biol Chem* 277, 5637-43.
Capell, A., Steiner, H., Willem, M., Kaiser, H., Meyer, C., Walter, J., Lammich, S., Multhaup, G. and Haass, C. (2000). Maturation and pro-peptide cleavage of beta-secretase. *J Biol Chem* 275, 30849-54.
Citron, M., Oltersdorf, T., Haass, C., McConlogue, L., Hung, A. Y., Seubert, P., Vigo-Pelfrey, C., Lieberburg, I. and Selkoe, D. J. (1992). Mutation of the beta-amyloid precursor protein in familial Alzheimer's disease increases beta-protein production. *Nature* 360, 672-4.
Cordy, J. M., Hussain, I., Dingwall, C., Hooper, N. M. and Turner, A. J. (2003). Exclusively targeting beta-secretase to lipid rafts by GPI-anchor addition up-regulates beta-site processing of the amyloid precursor protein. *Proc Natl Acad Sci U S A* 100, 11735-40.
Creemers, J. W., Ines Dominguez, D., Plets, E., Serneels, L., Taylor, N. A., Multhaup, G., Craessaerts, K., Annaert, W. and De Strooper, B. (2001). Processing of beta-secretase by furin and other members of the proprotein convertase family. *J Biol Chem* 276, 4211-7.

Dominguez, D., Tournoy, J., Hartmann, D., Huth, T., Cryns, K., Deforce, S., Serneels, L., Camacho, I. E., Marjaux, E., Craessaerts, K. et al. (2005). Phenotypic and biochemical analyses of BACE1- and BACE2-deficient mice. *J Biol Chem* 280, 30797-806.

Ehehalt, R., Keller, P., Haass, C., Thiele, C. and Simons, K. (2003). Amyloidogenic processing of the Alzheimer beta-amyloid precursor protein depends on lipid rafts. *J Cell Biol* 160, 113-23.

Farzan, M., Schnitzler, C. E., Vasilieva, N., Leung, D. and Choe, H. (2000). BACE2, a beta-secretase homolog, cleaves at the beta site and within the amyloid-beta region of the amyloid-beta precursor protein. *Proc Natl Acad Sci U S A* 97, 9712-7.

Fluhrer, R., Capell, A., Westmeyer, G., Willem, M., Hartung, B., Condron, M. M., Teplow, D. B., Haass, C. and Walter, J. (2002). A non-amyloidogenic function of BACE-2 in the secretory pathway. *J Neurochem* 81, 1011-20.

Fluhrer, R., Multhaup, G., Schlicksupp, A., Okochi, M., Takeda, M., Lammich, S., Willem, M., Westmeyer, G., Bode, W., Walter, J. et al., (2003). Identification of a beta-secretase activity, which truncates amyloid beta-peptide after its presenilin-dependent generation. *J Biol Chem* 278, 5531-8.

Goldsbury, C., Mocanu, M. M., Thies, E., Kaether, C., Haass, C., Keller, P., Biernat, J., Mandelkow, E. and Mandelkow, E. M. (2006). Inhibition of APP trafficking by tau protein does not increase the generation of amyloid-beta peptides. *Traffic* 7, 873-88.

Haass, C., Lemere, C. A., Capell, A., Citron, M., Seubert, P., Schenk, D., Lannfelt, L. and Selkoe, D. J. (1995). The Swedish mutation causes early-onset Alzheimer's disease by beta-secretase cleavage within the secretory pathway. *Nat Med* 1, 1291-6.

Haniu, M., Denis, P., Young, Y., Mendiaz, E. A., Fuller, J., Hui, J. O., Bennett, B. D., Kahn, S., Ross, S., Burgess, T. et al. (2000). Characterization of Alzheimer's beta-secretase protein BACE. A pepsin family member with unusual properties. *J Biol Chem* 275, 21099-106.

He, X., Chang, W. P., Koelsch, G. and Tang, J. (2002). Memapsin 2 (beta-secretase) cytosolic domain binds to the VHS domains of GGA1 and GGA2: implications on the endocytosis mechanism of memapsin 2. *FEBS Lett* 524, 183-7.

He, W., Lu, Y., Qahwash, I., Hu, X.Y., Chang, A. And Yan, R. (2004). Reticulon family members modulate BACE1 activity and amyloid-beta peptide generation. *Nat Med* 10, 959-65.

He, X., Li, F., Chang, W. P. and Tang, J. (2005). GGA proteins mediate the recycling pathway of memapsin 2 (BACE). *J Biol Chem* 280, 11696-703.

Holsinger, R. M., McLean, C. A., Beyreuther, K., Masters, C. L. and Evin, G. (2002). Increased expression of the amyloid precursor beta-secretase in Alzheimer's disease. *Ann Neurol* 51, 783-6.

Hu, X., Hicks, C. W., He, W., Wong, P., Macklin, W. B., Trapp, B. D. and Yan, R. (2006). Bace1 modulates myelination in the central and peripheral nervous system. *Nat Neurosci* 9, 1520-5.

Huse, J. T., Pijak, D. S., Leslie, G. J., Lee, V. M. and Doms, R. W. (2000). Maturation and endosomal targeting of beta-site amyloid precursor protein-cleaving enzyme. The Alzheimer's disease beta-secretase. *J Biol Chem* 275, 33729-37.

Hussain, I., Powell, D., Howlett, D. R., Tew, D. G., Meek, T. D., Chapman, C., Gloger, I. S., Murphy, K. E., Southan, C. D., Ryan, D. M. et al. (1999). Identification of a novel aspartic protease (Asp 2) as beta-secretase. *Mol Cell Neurosci* 14, 419-27.

Jacobsen, L., Madsen, P., Nielsen, M. S., Geraerts, W. P., Gliemann, J., Smit, A. B. and Petersen, C. M. (2002). The sorLA cytoplasmic domain interacts with GGA1 and -2 and defines minimum requirements for GGA binding. *FEBS Lett* 511, 155-8.

Kamal, A., Almenar-Queralt, A., LeBlanc, J. F., Roberts, E. A. and Goldstein, L. S. (2001). Kinesin-mediated axonal transport of a membrane compartment containing beta-secretase and presenilin-1 requires APP. *Nature* 414, 643-8.

Kamal, A., Stokin, G. B., Yang, Z., Xia, C. H. and Goldstein, L. S. (2000). Axonal transport of amyloid precursor protein is mediated by direct binding to the kinesin light chain subunit of kinesin-I. *Neuron* 28, 449-59.

Kametaka, S., Shibata, M., Moroe, K., Kanamori, S., Ohsawa, Y., Waguri, S., Sims, P. J., Emoto, K., Umeda, M. and Uchiyama, Y. (2003). Identification of phospholipid scramblase 1 as a novel interacting molecule with beta - secretase (beta -site amyloid precursor protein (APP) cleaving enzyme (BACE)). *J Biol Chem* 278, 15239-45.

Kitazume, S., Tachida, Y., Oka, R., Kotani, N., Ogawa, K., Suzuki, M., Dohmae, N., Takio, K., Saido, T. C. and Hashimoto, Y. (2003). Characterization of alpha 2,6-sialyltransferase cleavage by Alzheimer's beta -secretase (BACE1). *J Biol Chem* 278, 14865-71.

Koh, Y. H., von Arnim, C. A., Hyman, B. T., Tanzi, R. E. and Tesco, G. (2005). BACE is degraded via the lysosomal pathway. *J Biol Chem* 280, 32499-504.

Koo, E. H., Squazzo, S. L., Selkoe, D. J. and Koo, C. H. (1996). Trafficking of cell-surface amyloid beta-protein precursor. I. Secretion, endocytosis and recycling as detected by labeled monoclonal antibody. *J Cell Sci* 109 (Pt 5), 991-8.

Kuhn, P. H., Marjaux, E., Imhof, A., De Strooper, B., Haass, C. and Lichtenthaler, S. F. (2007). Regulated Intramembrane Proteolysis of the Interleukin-1 Receptor II by {alpha}-, beta-, and {gamma}-Secretase. *J Biol Chem* 282, 11982-95.

Laird, F. M., Cai, H., Savonenko, A. V., Farah, M. H., He, K., Melnikova, T., Wen, H., Chiang, H. C., Xu, G., Koliatsos, V. E. et al., (2005). BACE1, a major determinant of selective vulnerability of the brain to amyloid-beta amyloidogenesis, is essential for cognitive, emotional, and synaptic functions. *J Neurosci* 25, 11693-709.

Lazarov, O., Morfini, G. A., Lee, E. B., Farah, M. H., Szodorai, A., DeBoer, S. R., Koliatsos, V. E., Kins, S., Lee, V. M., Wong, P. C. et al., (2005). Axonal transport, amyloid precursor protein, kinesin-1, and the processing apparatus: revisited. *J Neurosci* 25, 2386-95.

Lee, E. B., Zhang, B., Liu, K., Greenbaum, E. A., Doms, R. W., Trojanowski, J. Q. and Lee, V. M. (2005). BACE overexpression alters the subcellular processing of APP and inhibits Abeta deposition in vivo. *J Cell Biol* 168, 291-302.

Li, Q. and Sudhof, T. C. (2004). Cleavage of amyloid-beta precursor protein and amyloid-beta precursor-like protein by BACE 1. *J Biol Chem* 279, 10542-50.

Lichtenthaler, S. F., Dominguez, D. I., Westmeyer, G. G., Reiss, K., Haass, C., Saftig, P., De Strooper, B. and Seed, B. (2003). The cell adhesion protein P-selectin glycoprotein ligand-1 is a substrate for the aspartyl protease BACE1. *J Biol Chem* 278, 48713-9.

Liu, K., Doms, R. W. and Lee, V. M. (2002). Glu11 site cleavage and N-terminally truncated A beta production upon BACE overexpression. *Biochemistry* 41, 3128-36.

Lo, A. C., Haass, C., Wagner, S. L., Teplow, D. B. and Sisodia, S. S. (1994). Metabolism of the "Swedish" amyloid precursor protein variant in Madin-Darby canine kidney cells. *J Biol Chem* 269, 30966-73.

Luo, Y., Bolon, B., Kahn, S., Bennett, B. D., Babu-Khan, S., Denis, P., Fan, W., Kha, H., Zhang, J., Gong, Y. et al., (2001). Mice deficient in BACE1, the Alzheimer's beta-secretase, have normal phenotype and abolished beta-amyloid generation. *Nat Neurosci* 4, 231-2.

Mathews, P. M., Guerra, C. B., Jiang, Y., Grbovic, O. M., Kao, B. H., Schmidt, S. D., Dinakar, R., Mercken, M., Hille-Rehfeld, A., Rohrer, J. et al.. (2002). Alzheimer's disease-related overexpression of the cation-dependent mannose 6-phosphate receptor increases Abeta secretion: role for altered lysosomal hydrolase distribution in beta-amyloidogenesis. *J Biol Chem* 277, 5299-307.

Offe, K., Dodson, S. E., Shoemaker, J. T., Fritz, J. J., Gearing, M., Levey, A. I. and Lah, J. J. (2006). The lipoprotein receptor LR11 regulates amyloid beta production and amyloid precursor protein traffic in endosomal compartments. *J Neurosci* 26, 1596-603.

Pastorino, L., Ikin, A. F., Nairn, A. C., Pursnani, A. and Buxbaum, J. D. (2002). The carboxyl-terminus of BACE contains a sorting signal that regulates BACE trafficking but not the formation of total A(beta). *Mol Cell Neurosci* 19, 175-85.

Perez, R. G., Squazzo, S. L. and Koo, E. H. (1996). Enhanced release of amyloid beta-protein from codon 670/671 "Swedish" mutant beta-amyloid precursor protein occurs in both secretory and endocytic pathways. *J Biol Chem* 271, 9100-7.

Puertollano, R., Aguilar, R. C., Gorshkova, I., Crouch, R. J. and Bonifacino, J. S. (2001). Sorting of mannose 6-phosphate receptors mediated by the GGAs. *Science* 292, 1712-6.

Qing, H., Zhou, W., Christensen, M. A., Sun, X., Tong, Y. and Song, W. (2004). Degradation of BACE by the ubiquitin-proteasome pathway. *Faseb J* 18, 1571-3.

Rentmeister, A., Bill, A., Wahle, T., Walter, J. and Famulok, M. (2006). RNA aptamers selectively modulate protein recruitment to the cytoplasmic domain of beta-secretase BACE1 in vitro. *Rna* 12, 1650-60.

Roberds, S. L., Anderson, J., Basi, G., Bienkowski, M. J., Branstetter, D. G., Chen, K. S., Freedman, S. B., Frigon, N. L., Games, D., Hu, K. et al. (2001). BACE knockout mice are healthy despite lacking the primary beta-secretase activity in brain: implications for Alzheimer's disease therapeutics. *Hum Mol Genet* 10, 1317-24.

Rockenstein, E., Mante, M., Alford, M., Adame, A., Crews, L., Hashimoto, M., Esposito, L., Mucke, L. and Masliah, E. (2005). High beta-secretase activity elicits neurodegeneration in transgenic mice despite reductions in amyloid-beta levels: implications for the treatment of Alzheimer disease. *J Biol Chem* 280, 32957-67.

Rogaeva, E., Meng, Y., Lee, J. H., Gu, Y., Kawarai, T., Zou, F., Katayama, T., Baldwin, C. T., Cheng, R., Hasegawa, H. et al. (2007). The neuronal sortilin-related receptor SORL1 is genetically associated with Alzheimer disease. *Nat Genet* 39, 168-177.

Saunders, A. M., Schmader, K., Breitner, J. C., Benson, M. D., Brown, W. T., Goldfarb, L., Goldgaber, D., Manwaring, M. G., Szymanski, M. H., McCown, N. et al., (1993). Apolipoprotein E epsilon 4 allele distributions in late-onset Alzheimer's disease and in other amyloid-forming diseases. *Lancet* 342, 710-1.

Schmechel, A., Strauss, M., Schlicksupp, A., Pipkorn, R., Haass, C., Bayer, T. A. and Multhaup, G. (2004). Human BACE forms dimers and colocalizes with APP. *J Biol Chem* 279, 39710-7.

Selkoe, D. J. (2001). Alzheimer's disease: genes, proteins, and therapy. *Physiol Rev* 81, 741-66.

Shi, X. P., Chen, E., Yin, K. C., Na, S., Garsky, V. M., Lai, M. T., Li, Y. M., Platchek, M., Register, R. B., Sardana, M. K. et al., (2001). The pro domain of beta-secretase does not confer strict zymogen-like properties but does assist proper folding of the protease domain. *J Biol Chem* 276, 10366-73.

Shi, X. P., Tugusheva, K., Bruce, J. E., Lucka, A., Wu, G. X., Chen-Dodson, E., Price, E., Li, Y., Xu, M., Huang, Q. et al., (2003). Beta-secretase cleavage at amino acid residue 34 in the amyloid beta peptide is dependent upon gamma-secretase activity. *J Biol Chem* 278, 21286-94.

Simons, K. and Toomre, D. (2000). Lipid rafts and signal transduction. Nat Rev *Mol Cell Biol* 1, 31-9.

Sinha, S., Anderson, J. P., Barbour, R., Basi, G. S., Caccavello, R., Davis, D., Doan, M., Dovey, H. F., Frigon, N., Hong, J. et al., (1999). Purification and cloning of amyloid precursor protein beta-secretase from human brain. *Nature* 402, 537-40.

Spoelgen, R., von Arnim, C. A., Thomas, A. V., Peltan, I. D., Koker, M., Deng, A., Irizarry, M. C., Andersen, O. M., Willnow, T. E. and Hyman, B. T. (2006). Interaction of the cytosolic domains of sorLA/LR11 with the amyloid precursor protein (APP) and beta-secretase beta-site APP-cleaving enzyme. *J Neurosci* 26, 418-28.

Steiner, H. and Haass, C. (2000). Intramembrane proteolysis by presenilins. *Nat Rev Mol Cell Biol* 1, 217-24.

Thinakaran, G., Teplow, D. B., Siman, R., Greenberg, B. and Sisodia, S. S. (1996). Metabolism of the "Swedish" amyloid precursor protein variant in neuro2a (N2a) cells. Evidence that cleavage at the "beta-secretase" site occurs in the golgi apparatus. *J Biol Chem* 271, 9390-7.

Valentine, J. S. and Hart, P. J. (2003). Misfolded CuZnSOD and amyotrophic lateral sclerosis. *Proc Natl Acad Sci U S A* 100, 3617-22.

Vassar, R. (2001). The beta-secretase, BACE: a prime drug target for Alzheimer's disease. *J Mol Neurosci* 17, 157-70.

Vassar, R., Bennett, B. D., Babu-Khan, S., Kahn, S., Mendiaz, E. A., Denis, P., Teplow, D. B., Ross, S., Amarante, P., Loeloff, R. et al., (1999). Beta-

secretase cleavage of Alzheimer's amyloid precursor protein by the transmembrane aspartic protease BACE. *Science* 286, 735-41.
von Arnim, C. A., Kinoshita, A., Peltan, I. D., Tangredi, M. M., Herl, L., Lee, B. M., Spoelgen, R., Hshieh, T. T., Ranganathan, S., Battey, F. D. et al., (2005). The low density lipoprotein receptor-related protein (LRP) is a novel beta-secretase (BACE1) substrate. *J Biol Chem* 280, 17777-85.
von Arnim, C. A., Spoelgen, R., Peltan, I. D., Deng, M., Courchesne, S., Koker, M., Matsui, T., Kowa, H., Lichtenthaler, S. F., Irizarry, M. C. et al., (2006). GGA1 acts as a spatial switch altering amyloid precursor protein trafficking and processing. *J Neurosci* 26, 9913-22.
von Arnim, C. A., Tangredi, M. M., Peltan, I. D., Lee, B. M., Irizarry, M. C., Kinoshita, A. and Hyman, B. T. (2004). Demonstration of BACE (beta-secretase) phosphorylation and its interaction with GGA1 in cells by fluorescence-lifetime imaging microscopy. *J Cell Sci* 117, 5437-45.
Wahle, T., Prager, K., Raffler, N., Haass, C., Famulok, M. and Walter, J. (2005). GGA proteins regulate retrograde transport of BACE1 from endosomes to the trans-Golgi network. *Mol Cell Neurosci* 29, 453-61.
Wahle, T., Thal, D. R., Sastre, M., Rentmeister, A., Bogdanovic, N., Famulok, M., Heneka, M. T. and Walter, J. (2006). GGA1 is expressed in the human brain and affects the generation of amyloid beta-peptide. *J Neurosci* 26, 12838-46.
Walter, J., Capell, A., Grunberg, J., Pesold, B., Schindzielorz, A., Prior, R., Podlisny, M. B., Fraser, P., Hyslop, P. S., Selkoe, D. J. et al., (1996). The Alzheimer's disease-associated presenilins are differentially phosphorylated proteins located predominantly within the endoplasmic reticulum. *Mol Med* 2, 673-91.
Walter, J., Fluhrer, R., Hartung, B., Willem, M., Kaether, C., Capell, A., Lammich, S., Multhaup, G. and Haass, C. (2001). Phosphorylation regulates intracellular trafficking of beta-secretase. *J Biol Chem* 276, 14634-41.
Walter, J., Kaether, C., Steiner, H. and Haass, C. (2001). The cell biology of Alzheimer's disease: uncovering the secrets of secretases. *Curr Opin Neurobiol* 11, 585-90.
Westmeyer, G. G., Willem, M., Lichtenthaler, S. F., Lurman, G., Multhaup, G., Assfalg-Machleidt, I., Reiss, K., Saftig, P. and Haass, C. (2004). Dimerization of beta-site beta-amyloid precursor protein-cleaving enzyme. *J Biol Chem* 279, 53205-12.
Willem, M., Garratt, A. N., Novak, B., Citron, M., Kaufmann, S., Rittger, A., DeStrooper, B., Saftig, P., Birchmeier, C. and Haass, C. (2006). Control of

peripheral nerve myelination by the beta-secretase BACE1. *Science* 314, 664-6.

Wong, H. K., Sakurai, T., Oyama, F., Kaneko, K., Wada, K., Miyazaki, H., Kurosawa, M., De Strooper, B., Saftig, P. and Nukina, N. (2005). beta Subunits of voltage-gated sodium channels are novel substrates of beta-site amyloid precursor protein-cleaving enzyme (BACE1) and gamma-secretase. *J Biol Chem* 280, 23009-17.

Wong, P. C., Waggoner, D., Subramaniam, J. R., Tessarollo, L., Bartnikas, T. B., Culotta, V. C., Price, D. L., Rothstein, J. and Gitlin, J. D. (2000). Copper chaperone for superoxide dismutase is essential to activate mammalian Cu/Zn superoxide dismutase. *Proc Natl Acad Sci U S A* 97, 2886-91.

Yamazaki, T., Koo, E. H. and Selkoe, D. J. (1996). Trafficking of cell-surface amyloid beta-protein precursor. II. Endocytosis, recycling and lysosomal targeting detected by immunolocalization. *J Cell Sci* 109 (Pt 5), 999-1008.

Yan, R., Bienkowski, M. J., Shuck, M. E., Miao, H., Tory, M. C., Pauley, A. M., Brashier, J. R., Stratman, N. C., Mathews, W. R., Buhl, A. E. et al., (1999). Membrane-anchored aspartyl protease with Alzheimer's disease beta-secretase activity. *Nature* 402, 533-7.

Yang, L. B., Lindholm, K., Yan, R., Citron, M., Xia, W., Yang, X. L., Beach, T., Sue, L., Wong, P., Price, D. et al., (2003). Elevated beta-secretase expression and enzymatic activity detected in sporadic Alzheimer disease. *Nat Med* 9, 3-4.

In: Cognitive Sciences Research Progress
Editor: Miao-Kun Sun

ISBN: 978-1-60456-392-4
© 2008 Nova Science Publishers, Inc.

Chapter 2

BRAIN VESICULAR MONOAMINE TRANSPORTER AND APOPTOSIS: PERSPECTIVES ON DEVELOPMENT AND NEURODEGENERATION

Léa Stankovski[1], Patricia Gaspar[1] and Olivier Cases[1,2]

[1] INSERM, UMR616, Université Pierre et Marie Curie, Hôpital de la Pitié-Salpétrière, 47 Boulevard de l'Hôpital, 75013 Paris, France.
[2] INSERM, UMR676, Université Denis Diderot, Hôpital Robert Debré, 48 Boulevard Sérurier, 75019 Paris, France.

ABSTRACT

The neuronal isoform of vesicular monoamine transporter, VMAT2, is responsible for packaging dopamine, norepinephrine, and serotonin into synaptic vesicles and thereby plays an essential role in monoamine neurotransmission. This sequestering action is important for normal synaptic release of monoamines but it may also act to keep intracellular levels of the monoamine transmitters below potentially toxic levels. During embryonic and postnatal development, VMAT2 is expressed in many non-aminergic neurons in the brain, long before synapses are formed, suggesting that it could have non-synaptic roles. Here, we review recent evidences indicating a role for VMAT2 in the control of developmental cell death in the cerebral cortex and in models of Parkinson related disorders. Abnormalities in VMAT2 functions have been suggested to play a key role in the etiology of a number of disorders, including Parkinson's disease and addiction.

Key words: vesicular monoamine transporter 2, monoamines, apoptosis, cerebral cortex, Parkinson's disease.

INTRODUCTION

Vesicular neurotransmitter transporters are essential components of secretory vesicle, which are responsible for storage and regulated secretion of neurotransmitters in neurons and neuroendocrine cells. The nature of the vesicular transporters expressed by a given cell dictates what kind, and how much neurotransmitter is released during synaptic transmission (Eiden, 2000). Vesicular transporters are members of the toxin-extruding proton-translocating antiporter family. As such, another possibly important function of these transporters is to protect the cells from accumulating potentially toxic levels of transmitters intracellularly.

The vesicular monoamine transporters (VMAT) are responsible for the translocation of biogenic monoamines (serotonin, dopamine, norepinephrine, and histamine) from the cytoplasm into synaptic vesicles via a proton electrochemical gradient generated by the vacuolar type H+-adenosine triphosphate. Two vesicular monoamine transporters have been identified by expression, cloning, and pharmacology (Erickson et al., 1992; Liu et al., 1992; Erickson and Eiden, 1993). VMAT1 and VMAT2 arise from two separate genes, yet have extensive homology.

Only VMAT1 is expressed in serotonin-accumulating enterochromaffin cells, whereas only VMAT2 is found in enterochromaffin-like histaminocytes of the stomach (Eiden, 2000). Differences in the substrate recognition and inhibitor sensitivities between VMAT1 and VMAT2 have been studied, using membrane vesicles prepared from stable transformed cell lines from Chinese hamster ovaries (CHO) that express the respective proteins (Peter et al., 1994). The major difference between the two transporters is that VMAT2 efficiently transports histamine and VMAT1 does not. VMAT1 may have evolved during the evolutionary emergence of histamine as a secreted effector molecule, restricting the storage of histamine in enterochromaffin cells of the lower gut (Hoffman et al., 1998).

Both VMAT1 and VMAT2 are more widely expressed in the brain during embryonic and early postnatal development .

VMAT2 VERSUS MONOAMINE OXIDASE

Oxidative deamination of monoamines by monoamine oxidase type A (MAOA) and type B (MAOB) is accompanied by the reduction of molecular oxygen to a toxic product, hydrogen peroxide (Shih et al., 1999). Therefore, maintenance of low cytoplasmic concentrations of neurotransmitters by their reuptake into synaptic vesicles is important to minimize their inherent toxicity. Furthermore, storage of neurotransmitters in synaptic vesicles precludes their metabolism in the cytoplasmic compartment and reduces the synthetic demands on the cell. In the central nervous system, VMAT2 is the only transporter that moves cytoplasmic monoamines into synaptic vesicles for storage and subsequent exocytotic release. MAOA has higher affinity for the substrates serotonin (5-HT), norepinephrine (NE), dopamine (DA), histamine (HA), and the inhibitor clorgyline, whereas MAO B has higher affinity for phenylethylamine (PEA), benzylamine, and the inhibitor deprenyl.

MAOA and MAOB display non-overlapping patterns of mRNA and protein expression. MAOA is abundantly expressed in noradrenergic, and dopaminergic neurons and restricted neuronal populations. Conversely, MAOB expression is highly expressed in serotoninergic and histaminergic neurons, as well as in astrocytes (Vitalis et al., 2002).

DEVELOPMENTAL EXPRESSION OF VMAT2

In all the norepinephrine-, epinephrine-, serotonin-, dopamine-, and histamine-containing cell groups, VMAT2 is expressed one day after embryonic differentiation; i.e., E13-E14 for serotoninergic neurons of the rat dorsal raphe (Hansson et al., 1998a,b). In addition, VMAT2 is expressed in selective non-monoaminergic neuronal populations during embryonic and postnatal development. These neurons are not enzymatically equipped to produce amines and instead appear to have a glutamatergic phenotype. VMAT2 is expressed during a defined period in each neuronal population. For instance, the glutamatergic neurons of the sensory thalamus express VMAT2 from E15 to P15, whereas in the cerebral cortex, VMAT2 is expressed from E18 to P5 (Hansson et al., 1998a,b, Lebrand et al., 1998). In most neuronal populations, VMAT2 is co-expressed with the plasma membrane serotonin transporter (Lebrand et al., 1998), suggesting that it is capable of handling serotonin. Curiously, a few populations expressing VMAT2 do not display other known monoaminergic markers.

However, all the VMAT2-expressing neurons are capable of accumulating serotonin, under the conditions where the extracellular levels of this amine are raised (Cases et al., 1998).

MICE LACKING VMAT2

Genetic models, such as those afforded by the VMAT2 deficient mice (Fon et al., 1997; Takahashi et al., 1997; Wang et al., 1997) allow the re-evaluation of the effects of VMAT2 depletion without the inherent limitations of pharmacological treatments. Homozygous VMAT2 KO mice are severely atrophic and most pups die by P4 (Fon et al., 1997; Takahashi et al., 1997; Wang et al., 1997). VMAT2 KO pups are born in the expected Mendelian ratio and their gross brain morphology appears normal. In addition, monoamine cell groups in the brainstem and their projections have a normal appearance. In particular, DA islands and the patch-matrix system appear unaffected by the elimination of VMAT2, suggesting that monoamine release does not have a role in the formation of these connections (Fon et al., 1997). Nonetheless, VMAT2 KO pups move little and feed poorly compared to their wild-type and heterozygous littermates, accounting for their growth failure and early postnatal mortality. Consistent with the role for VMAT2, DA neurons cultured from the VMAT2 KO as well as striatal slices prepared from these animals show no depolarization-evoked DA release (Fon et al., 1997; Wang et al., 1997). However, amphetamine still induces DA release from VMAT2 KO neurons, indicating that VMAT2 is not required for the action of the psychostimulant, at least under these circumstances where cytoplasmic DA may already be elevated. Indeed, administration of amphetamine to VMAT2 KO increases movements, such as walking and feeding, and enables the animals to survive for several weeks (Fon et al., 1997). Thus, precisely regulated exocytotic release of monoamines does not seem to be required for several complex behaviours.

Disruption of the VMAT2 gene results in dramatically reduced brain levels of DA, NE, and 5-HT (Fon et al., 1997; Wang et al., 1997). This could be a consequence of reduced monoamine synthesis or increased degradation. However, disruption of VMAT2 does not alter the expression level of the rate-limiting catecholamine biosynthetic enzyme tyrosine hydroxylase and TH activity appears to be upregulated (Wang et al., 1997). Furthermore, monoamine metabolites are not reduced in the VMAT2 KO mice (Alvarez et al., 2002). All these observations suggest that VMAT2 ablation causes an increase in monoamine catabolism rather that a decrease in synthesis. Interestingly, an administration of the MAOA inhibitor clorgyline increases 5-HT levels, supporting this interpretation. However, clorgyline has no effect on DA levels, indicating that additional mechanisms act to control the level of this neurotransmitter (Fon et al., 1997). In VMAT2-MAOA DKO mouse, pups are severely hypomorphic in comparison to the wild-type or MAOA KO mice but they survive until P13 (Alvarez et al., 2002).

VMAT2 AND PROGRAMMED CELL DEATH IN THE CEREBRAL CORTEX

The development of the cerebral cortex is a sequential process that includes programmed cell death (PCD). Approximately half of the neurons produced during corticogenesis are thought to die (Ferrer et al., 1992; Blaschke et al., 1996). Two consecutive waves of PCD affect the cortical neurons at different periods of their development. The first wave consists of cell death of the proliferating precursors in the ventricular and subventricular zones, which appears to be closely linked to cell cycle regulation (Thomaidou et al., 1997). The second wave affects postmitotic neurons at later stages and may be involved in matching the size of neuronal populations to that of their targets during the formation and the maintenance of synapses (Ferrer et al., 1992). At both periods, cell death is apoptotic and the molecular machinery relies on the mitochondrial pathways of intracellular signal transduction (Dikranian et al., 2001).

Using this model, we found a significant increase of cell death in the supragranular layers of the cingulate and retrosplenial areas (Skankovski et al., 2007). This effect is regionally specific since some regions such as the primary somatosensory cortex show no changes (Alvarez et al., 2002), whereas other regions, such as the hippocampus or striatum, display an increase in cell death (unpublished). Electron microscopic analysis and TUNEL staining confirmed that this was related to an increase in apoptosis. Additional characterization indicated an increase in caspase-9-, and caspase-3-immunoreactive profiles, suggesting that the increase in cell death observed in the cortex of VMAT2 KO mice is attributable to an increase of apoptotic caspase-dependent mechanism. We also observed a specific decrease in the mRNA levels of the anti-apoptotic factor Bcl-XL, whereas mRNA levels of the proapoptotic factor Bax remained unchanged. This suggests that the decrease of the Bcl-XL/Bax mRNA ratio is probably the main phenomenon at play in the increased cell death in VMAT2 KO cerebral cortex (Stankovski et al., 2007). However, the precise mechanisms by which the depletion of VMAT2 leads to this change in gene expression and to exaggerated cell death remain elusive at the moment.

In synaptic vesicles, monoamines are protected from degradation by MAO activity. MAOA is the major form expressed in specific neuronal populations, such as the serotonergic neurons during early postnatal life (Vitalis et al., 2002). Thus, invalidating MAOA in a VMAT2 KO genetic background should maintain a pool of cytoplasmic 5-HT by preventing its degradation. Indeed, high serotonin levels were detected in the brain of VMAT2-MAOA DKO mice, whereas dopamine and norepinephrine levels remained low or undetectable (Alvarez et al., 2002). Interestingly, the increase in cell death observed in the VMAT2 KO was reversed in the cerebral cortex of VMAT2-MAOA DKO, indicating that increases in brain serotonin levels could have a neuroprotectant effect on cortical neurons (Stankovski et al., 2007). This effect is in agreement with a previous report by Persico et al., (2003) in the brain of mice lacking 5-HTT. 5-HTT KO mice display high serotonin levels and reduced PCD, as judged by the decrease of TUNEL-positive cells in their brain. We found that Bcl-XL mRNA levels returned to normal in the cerebral cortex of VMAT2-MAOA DKO, and no decrease in expression of the proapoptotic factor Bax was noted (Stankovski et al., 2007). This suggests that high serotonin levels maintain a normal Bcl-XL/Bax mRNA ratio. As a consequence, a normal level of cell death is observed in the cerebral cortex of VMAT2-MAOA DKO. The mechanisms that allow this normalization at both the cellular and molecular levels are not fully understood but likely involve one or several 5-HT receptor subtypes. We found that 5-HT2 receptor activation led to a partial restoration of the increased cell death phenotype in VMAT2 KO (Stankovski et al., 2007). These results are consistent with the prosurvival activity of 5-HT2 receptor activation that was noted previously in vitro (Dooley et al., 1997). Because activation of 5-HT2A/2C receptors leads to an augmentation of the trophic factor, brain-derived neurotrophic factor (BDNF) (Vaidya et al., 1997) and mice lacking the catalytic domain of trkB, the high-affinity receptor of BDNF, display an increase in cell death in the cingulate cortex (Alcantara et al., 1997), we hypothesized that trkB-dependent mechanisms could be involved in the anti-apoptotic cascade triggered by high serotonin levels. However, quantification of BDNF mRNA levels showed no differences between normal, VMAT2 KO, or VMAT2-MAOA DKO in the cingulate cortex. Furthermore, trkB KO showed a rescue of cell death in conditions of increased 5-HT levels (Stankovski et al., 2007). These results suggest that cell death in the VMAT2 KO is not attributable to a decrease of trkB signalling. However, we cannot rule out signalling via p75 neurotrophin receptor (Kalb, 2005).

PARKINSON'S DISEASE

Parkinson's disease is a progressively degenerative disorder that affects preferentially neurons of the substantia nigra. In this regard, DA may play a role as an endogenous toxin, since the normal metabolism of DA produces hydrogen peroxide as a by-product (Adams et al., 2001). Accordingly, pharmacological enhancement of DA sequestration by VMAT2, to prevent oxidation of DA in the cytoplasm, could be an interesting strategy for treatment of Parkinson's disease.

Exposure to the neurotoxin N-methyl-4-phenyltetrahydropyridine (MPTP) results in clinical symptoms that resemble Parkinson's disease (Langston, 1996). N-Methyl-4-phenylpyridinium (MPP+), the active toxic metabolite of MPTP, is a substrate for VMAT2 (Scherman et al., 1988; Moriyama et al., 1993). VMAT2 sequesters MPP+ in synaptic vesicles and thereby protects catecholamine-containing neurons from MPP+-induced toxicity and degeneration (Takahashi et al., 1997; Speciale et al., 1998; German et al., 2000; Staal et al., 2000). Studies using heterozygous VMAT2 knockout mice show that the knockouts are more susceptible to the neurotoxic effects of MPTP, compared with the wild-type mice (Takahashi et al., 1997; Gainetdinov et al., 1998; Mooslehner et al., 2001). Furthermore, heterozygous VMAT2 knockout mice are more sensitive to methamphetamine-induced neurotoxicity and are more vulnerable to the toxic effects of L-3,4-dihydroxyphenylalanine (L-DOPA, a DA precursor used to treat Parkinson's disease), compared with wild-type mice (Fumagalli et al., 1999; Kariya et al., 2005). The latter results suggest that reduction in VMAT2 activity might attenuate the efficacy of L-DOPA therapy in Parkinson's patients. Finally, increased sequestration of DA in synaptic vesicles by VMAT2 has been suggested to be protective in Parkinson's disease (Glatt et al., 2006).

Taken together, the results of the above studies indicate that VMAT2 expression and function are important in counteracting the neurotoxicity of MPP+ and perhaps of other environmental and endogenous neurotoxins that play an etiologic role in neurodegenerative diseases.

SPONSORS

This work was supported by Institut National de la Santé et de la Recherche Médicale, Centre National de la Recherche Scientifique (O.C.), University Paris 6, and Agence Nationale de la Recherche (-05- APV05146DSA).

REFERENCES

Alcantara S, Frisen J, del Rio JA, Soriano E, Barbacid M, Silos-Santiago I (1997) TrkB signaling is required for postnatal survival of CNS neurons and protects hippocampal and motor neurons from axotomy-induced cell death. *J Neurosci* 17:3623-3633.

Alvarez C, Vitalis T, Fon EA, Hanoun N, Hamon M, Seif I, Edwards R, Gaspar P, Cases O (2002) Effects of genetic depletion of monoamines on somatosensory cortical development. *Neuroscience* 115:753-764.

Blaschke AJ, Staley K, Chun J (1996) Widespread programmed cell death in proliferative and postmitotic regions of the fetal cerebral cortex. *Development* 122:1165-1174.

Cases O, Lebrand C, Giros B, Vitalis T, De Maeyer E, Caron MG, Price DJ, Gaspar P, Seif I (1998) Plasma membrane transporters of serotonin, dopamine, and norepinephrine mediate serotonin accumulation in atypical locations in the developing brain of monoamine oxidase A knock-outs. *J Neurosci* 18: 6914-6927.

Dikranian K, Ishimaru MJ, Tenkova T, Labruyere J, Qin YQ, Ikonomidou C, Olney JW (2001) Apoptosis in the in vivo mammalian forebrain. *Neurobiol Dis* 8:359-379.

Dooley AE, Pappas IS, Parnavelas JG (1997) Serotonin promotes the survival of cortical glutamatergic neurons in vitro. *Exp Neurol* 148:205-214.

Erickson JD, Eiden LE (1993) Functional identification and molecular cloning of a human brain vesicle monoamine transporter. *J Neurochem* 61:2314-2317.

Erickson JD, Eiden LE, Hoffman BJ (1992) Expression cloning of a reserpine-sensitive vesicular monoamine transporter. *Proc Natl Acad Sci U S A* 89:10993-10997.

Ferrer I, Soriano E, del Rio JA, Alcantara S, Auladell C (1992) Cell death and removal in the cerebral cortex during development. *Prog Neurobiol* 39:1-43.

Fon EA, Pothos EN, Sun BC, Killeen N, Sulzer D, Edwards RH (1997) Vesicular transport regulates monoamine storage and release but is not essential for amphetamine action. *Neuron* 19:1271-1283.

Fumagalli F, Gainetdinov RR, Wang YM, Valenzano KJ, Miller GW, Caron MG (1999) Increased methamphetamine neurotoxicity in heterozygous vesicular monoamine transporter 2 knock-out mice. *J Neurosci* 19:2424-2431.

Gainetdinov RR, Fumagalli F, Wang YM, Jones SR, Levey AI, Miller GW, Caron MG (1998) Increased MPTP neurotoxicity in vesicular monoamine transporter 2 heterozygote knockout mice. *J Neurochem* 70:1973-1978.

German DC, Liang CL, Manaye KF, Lane K, Sonsalla PK (2000) Pharmacological inactivation of the vesicular monoamine transporter can enhance 1-methyl-4-phenyl-1,2,3,6-tetrahydropyridine-induced neurodegeneration of midbrain dopaminergic neurons, but not locus coeruleus noradrenergic neurons. *Neuroscience* 101:1063-1069.

Glatt CE, Wahner AD, White DJ, Ruiz-Linares A, Ritz B (2006) Gain-of-function haplotypes in the vesicular monoamine transporter promoter are protective for Parkinson disease in women. *Hum Mol Genet* 15:299-305.

Hansson SR, Hoffman BJ, Mezey E (1998a) Ontogeny of vesicular monoamine transporter mRNAs VMAT1 and VMAT2. I The developing rat central nervous system. *Dev Brain Res* 110: 135-158.

Hansson SR, Mezey E, Hoffman BJ (1998b) Ontogeny of vesicular monoamine transporter mRNAs VMAT1 and VMAT2. II Expression in neural crest derivatives and their target sites in the rat. *Dev Brain Res* 110: 159-174.

Hoffman BJ, Hansson SR, Mezey E, Palkovits M (1998) Localization and dynamic regulation of biogenic amine transporters in the mammalian central nervous system. *Front Neuroendocrinol* 19:187-231.

Kalb R (2005) The protean actions of neurotrophins and their receptors on the life and death of neurons. *Trends Neurosci* 28:5-11.

Kariya S, Takahashi N, Hirano M, Ueno S (2005) Increased vulnerability to L-DOPA toxicity in dopaminergic neurons From VMAT2 heterozygote knockout mice. *J Mol Neurosci* 27:277-279.

Langston JW (1996) The etiology of Parkinson's disease with emphasis on the MPTP story. *Neurology* 47:S153-160.

Lebrand C, Cases O, Wehrle R, Blakely RD, Edwards RH, Gaspar P (1998) Transient developmental expression of monoamine transporters in the rodent forebrain. *J Comp Neurol* 401:506-524.

Liu Y, Peter D, Roghani A, Schuldiner S, Prive GG, Eisenberg D, Brecha N, Edwards RH (1992) A cDNA that suppresses MPP+ toxicity encodes a vesicular amine transporter. *Cell* 70:539-551.

Mooslehner KA, Chan PM, Xu W, Liu L, Smadja C, Humby T, Allen ND, Wilkinson LS, Emson PC (2001) Mice with very low expression of the vesicular monoamine transporter 2 gene survive into adulthood: potential mouse model for parkinsonism. *Mol Cell Biol* 21:5321-5331.

Moriyama Y, Amakatsu K, Futai M (1993) Uptake of the neurotoxin, 4-methylphenylpyridinium, into chromaffin granules and synaptic vesicles: a proton gradient drives its uptake through monoamine transporter. *Arch Biochem Biophys* 305:271-277.

Persico AM, Baldi A, Dell'Acqua ML, Moessner R, Murphy DL, Lesch KP, Keller F (2003) Reduced programmed cell death in brains of serotonin transporter knockout mice. *Neuroreport* 14:341-344.

Peter D, Jimenez J, Liu Y, Kim J, Edwards RH (1994) The chromaffin granule and synaptic vesicle amine transporters differ in substrate recognition and sensitivity to inhibitors. *J Biol Chem* 269:7231-7237.

Scherman D, Darchen F, Desnos C, Henry JP (1988) 1-Methyl-4-phenylpyridinium is a substrate of the vesicular monoamine uptake system of chromaffin granules. *Eur J Pharmacol* 146:359-360.

Shih JC, Chen K, Ridd MJ (1999) Monoamine oxidase: from genes to behavior. *Annu Rev Neurosci* 22:197-217.

Speciale SG, Liang CL, Sonsalla PK, Edwards RH, German DC (1998) The neurotoxin 1-methyl-4-phenylpyridinium is sequestered within neurons that contain the vesicular monoamine transporter. *Neuroscience* 84:1177-1185.

Staal RG, Hogan KA, Liang CL, German DC, Sonsalla PK (2000) In vitro studies of striatal vesicles containing the vesicular monoamine transporter (VMAT2): rat versus mouse differences in sequestration of 1-methyl-4-phenylpyridinium. *J Pharmacol Exp Ther* 293:329-335.

Stankovski L, Alvarez C, Ouimet T, Vitalis T, El-Hachimi KH, Price D, Deneris E, Gaspar P, Cases O (2007) Developmental cell death is enhanced in the cerebral cortex of mice lacking the brain vesicular monoamine transporter. *J Neurosci* 27:1315-1324.

Takahashi N, Miner LL, Sora I, Ujike H, Revay RS, Kostic V, Jackson-Lewis V, Przedborski S, Uhl GR (1997) VMAT2 knockout mice: heterozygotes display reduced amphetamine-conditioned reward, enhanced amphetamine locomotion, and enhanced MPTP toxicity. *Proc Natl Acad Sci U S A* 94:9938-9943.

Thomaidou D, Mione MC, Cavanagh JF, Parnavelas JG (1997) Apoptosis and its relation to the cell cycle in the developing cerebral cortex. *J Neurosci* 17:1075-1085.

Vaidya VA, Marek GJ, Aghajanian GK, Duman RS (1997) 5-HT2A receptor-mediated regulation of brain-derived neurotrophic factor mRNA in the hippocampus and the neocortex. *J Neurosci* 17:2785-2795.

Vitalis T, Fouquet C, Alvarez C, Seif I, Price D, Gaspar P, Cases O (2002) Developmental expression of monoamine oxidases A and B in the central and peripheral nervous systems of the mouse. *J Comp Neurol* 442:331-347.

Wang YM, Gainetdinov RR, Fumagalli F, Xu F, Jones SR, Bock CB, Miller GW, Wightman RM, Caron MG (1997) Knockout of the vesicular monoamine transporter 2 gene results in neonatal death and supersensitivity to cocaine and amphetamine. *Neuron* 19:1285-1296.

In: Cognitive Sciences Research Progress ISBN: 978-1-60456-392-4
Editor: Miao-Kun Sun © 2008 Nova Science Publishers, Inc.

Chapter 3

MILD COGNITIVE IMPAIRMENT IS TOO LATE: THE CASE FOR PRESYMPTOMATIC DETECTION AND TREATMENT OF ALZHEIMER'S DISEASE

Charles D. Smith

Department of Neurology & Alzheimer's Disease Center, University of Kentucky College of Medicine, Lexington, KY

ABSTRACT

The thesis of this review is that the earliest cognitive symptoms in dementia represent the exhaustion of compensatory mechanisms in the brain which counteract underlying Alzheimer's disease (AD), vascular, Lewy body, and other neuropathology. These pathologies may accumulate gradually over years and perhaps enter a phase of acceleration as compensatory resources are outstripped. Evidence for this model of a prolonged presymptomatic period in AD has come from recent imaging, neuropathologic, and basic science studies. These studies and the potential consequences for diagnosis and treatment are presented. The conclusion is that using onset symptoms as the signal to begin disease-modifying treatment for AD is too late; this treatment must begin earlier, before symptoms begin, to preserve brain function. Therefore presymptomatic detection is a critical research goal in AD.

Key words: Alzheimer's disease, prevention, review, human studies, mild cognitive impairment, neural plasticity

1. INTRODUCTION

Presymptomatic detection is different from the concept of detecting cognitive or clinical abnormalities prior to the diagnosis of dementia, or even the very earliest symptoms that presage dementia years later. It refers to detecting the presence of an underlying pathologic process that later expresses itself as a dementia when persons are manifestly normal. The many reasons to suspect this is possible are described here.

This is not an unbiased and dispassionate survey of literature on this topic. It has a specific viewpoint. The thesis is that the earliest cognitive symptoms in dementia represent the exhaustion of compensatory mechanisms in the brain which counteract accumulating Alzheimer's disease (AD), vascular, Lewy body, and other neuropathology. Detecting the presence of an underlying asymptomatic pathology may be indirect, e.g., one may be detecting brain compensations for pathology that would otherwise be unobservable. Though indirect, it is a form of presymptomatic detection.

The motivation for addressing presymptomatic detection now is the imminent availability of disease-modifying treatments for AD. Detecting threshold amounts of AD pathology before symptoms appear offers the hope that symptoms could be delayed or even prevented by applying these treatments in persons with this pathology. This threshold may be determined by several contributions to brain reserve discussed in this article. A summary argument for presymptomatic detection and treatment is presented in the conclusion.

2. THE BRAIN RESERVE HYPOTHESIS

The brain reserve hypothesis is the theoretical construct that inherited, developmental and cultural factors influence the symptomatic expression of specific underlying brain pathologies. Examples of such neuropathologies include human immunodeficiency virus (HIV) encephalopathy (Stern, Silva, Chaisson, & Evans, 1996) and neurodegenerative diseases, e.g., Alzheimer's disease (Mortimer, 1997; Mortimer, Borenstein, Gosche, & Snowdon, 2005; Mosconi et al., 2005; Wolf, Julin, Gertz, Winblad, & Wahlund, 2004) and frontotemporal dementia (Perneczky, Diehl-Schmid, Drzezga, & Kurz, 2007; Perneczky, Diehl-Schmid, Pohl, Drzezga, & Kurz, 2007). Examples of factors constituting reserve are years of education (McDowell, Xi, Lindsay, & Tierney, 2007), socioeconomic status level (Stern et al., 1994), head circumference (Borenstein Graves et al., 2001), and intelligence quotient. The neurobiologic basis of each these factors' contribution to reserve is often unstated or heuristically justified in the absence of scientific data other than that from studies demonstrating association between the factor and symptom expression. Nonetheless it is quite plausible that intrinsic or acquired brain properties may confer resistance to the effects of neuropathology, and therefore modify whether and when that neuropathology is expressed as symptoms, changes in cognitive performance or behavior.

2.1. Normal Developmental Changes in the Brain during Childhood and Adolescence

Here the focus is on the key findings related to the concept of brain reserve as reflected in changes in brain anatomy and histology, tissue density, diffusivity, and functional characteristics from conception to early adulthood.

In utero measurements by magnetic resonance imaging (MRI) and ultrasound demonstrate rapid, approximately linear, growth of the brain from 10 to 40 weeks gestation (Figure 1), corresponding to neurogenesis, and later in the second and third trimesters, increasing axonal sprouting, axonal route-finding, and synaptogenesis. As an illustration of underlying cortical histology during this period, the complex but ordered development of the human hippocampus proceeds from the primordial zonal arrangement (the ventricular neurogenesis source and intermediate zones, the neuronal migration target cortical plate, and outer neuron-poor marginal zone) (Noctor, Martinez-Cerdeno, & Kriegstein, 2007) at 9 weeks, through elaboration of subfields at 15-19 weeks, to morphologic maturity at 32-34 weeks gestation (Arnold & Trojanowski, 1996). The cytoarchitecture of the hippocampus is maintained after this time, although decreased neuronal density and enlargement of neurons occur throughout adolescence and beyond. Asynchrony of hippocampal complex development occurs, with the subiculum reaching maturity earliest, and the dentate latest in morphologic maturity. Myelenation only proceeds in this region in the weeks before birth, and continues through adolescence. This principal sequence of events and regional asynchrony is representative of other human cortical areas during brain development (Sidman & Rakic, 1982).

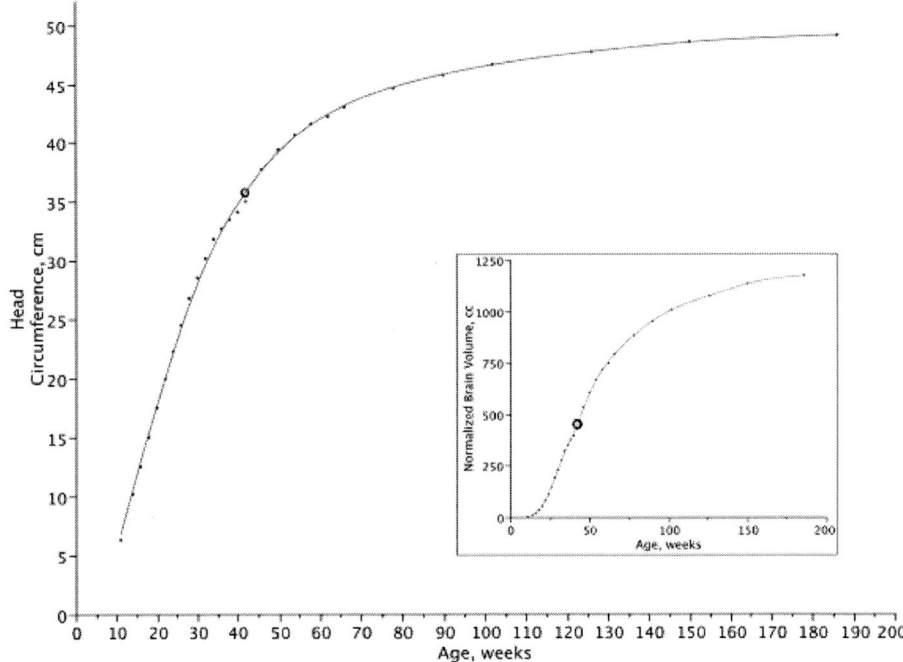

Fig. 1. Head circumference (HC) in centimeters from 10 weeks in utero (from Lessoway, Schulzer, Wittmann, Gagnon, & Wilson, (2008)) to 3 postnatal years, in weeks (from CDC growth charts, http://www.cdc.gov/nchs, 2007). There is approximately linear HC growth to at least 36 weeks gestation, corresponding in large part to neurogenesis and neuronal migration, and gradually diminishing postnatal HC growth associated with synaptogenesis and myelination processes. For continuity, the postnatal HC is averaged between boys and girls because the in utero results were not reported by gender. The curve is a spline smoothing fit to the 50th percentile data to show the trend; the apparent discontinuity at term could be due to a true slowing of head growth in the weeks just before birth, or differences in definition of gestational age, gender distribution, and technique of HC measurement between sources. Inset shows total volume (TIV) change estimated from HC, normalized by average TIV at term (427.4 cc;(Zacharia et al., 2006)). The TIV increases by approximately 11% between 3 and 16 years of age (Zhang et al., 2005). Circle (o) marks weeks at birth.

Only recently have brain imaging techniques allowed measurements of gray matter (GM) and white matter (WM) alterations with normal postnatal development. Gray matter density, used as a proxy for cortical developmental maturation, shows an increase before puberty followed by a decrease in adolescence and early adulthood that can be interpreted as a "pruning" process associated with reorganization and consolidation of efficient synaptic connections

(Courchesne et al., 2000; Giedd et al., 1999; Jernigan & Tallal, 1990; Jernigan, Trauner, Hesselink, & Tallal (1991); Sowell, Thompson, Tessner, & Toga, 2001; Wilke, Krageloh-Mann, & Holland, 2007). This pruning process is invisible to head circumference measurements (Figure 1) because head circumference reflects maximum achieved brain size and would not decrease despite a small decrease with brain volume during adolescence. Cortical GM density maps based on longitudinal MRI data from normal subjects in the age range 4 – 21 years demonstrate early maturation (4 – 10 years) in the limbic and primary sensorimotor cortical regions, and later maturation (10 – 20 years) in modality-specific association and polymodal neocortex (Figure 2) (Gogtay et al., 2004). It is notable that even the primary regions showing the earliest and greatest reduction in GM density still retain the capacity for synaptic reorganization into adulthood (Ramachandran, 2005a, b).

Normal WM development in childhood and adolescence follows the sequence classically described by Yakovlev and Lecours (Yakovlev & Lecours, 1967), later observed using MRI (Barkovich, Kjos, Jackson & Norman, 1988). Myelination in the CNS proceeds from early maturation in basic motor and sensory systems in the brainstem, and begins initially in the corticospinal and primary sensory tracts and the optic tracts and radiations in the cerebral hemispheres. Myelination is synchronous with the cortical maturation pattern described earlier, with later stages of maturation occurring in modality-specific association and polymodal neocortex, particularly in the frontal lobe U-fibers. Association bundles, e.g., the corpus callosum, increase in relative size during the first 20 years of life, paralleling these later changes (Giedd et al., 1999; Rauch & Jinkins, 1994). Diffusion tensor MRI studies have confirmed and refined these observations (Schmithorst, Wilke, Dardzinski, & Holland, 2002), for example demonstrating regional gender-specific effects: females aged 5–18 years demonstrated greater age-related increases in fiber density (by mean diffusivity) than males in associative regions, whereas males had greater absolute organization of myelinated tracts (by fractional anisotropy) in the same regions (Schmithorst, Holland, & Dardzinski, 2007). The myelination process continues well into adulthood, for example in the hippocampal/parahippocampal region (Benes, Turtle, Khan, & Farol, 1994).

Functional studies to complement regional GM and WM developmental patterns have focused on language development. Word fluency tasks activate more widespread cortical areas, e.g., the right inferior frontal lobe in 11-year old children compared to 28 year-old adults. Language lateralization to a verb generation task likewise increases between 5 and 20 years of age (Szaflarski, Holland, Schmithorst, & Byars, 2006). In word-picture matching over a similar

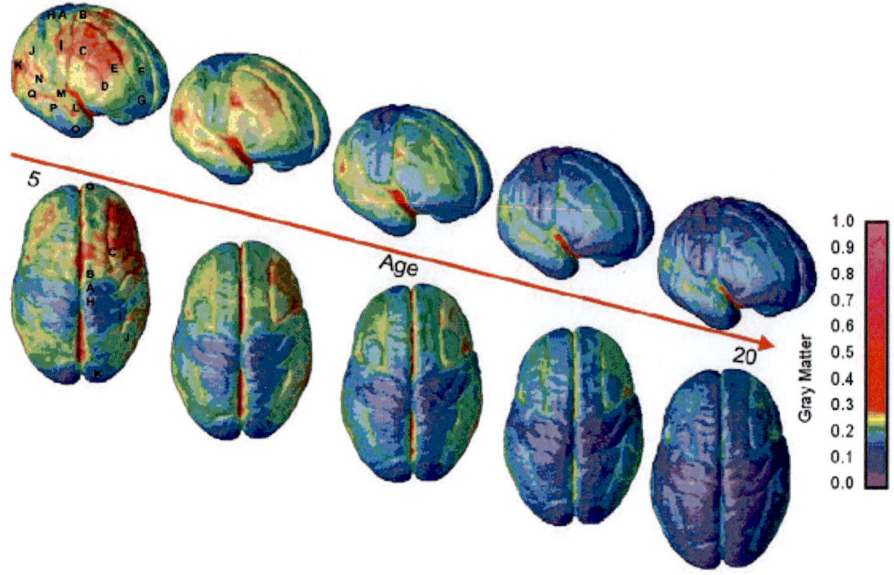

Fig. 2. Illustration from (Gogtay et al., 2004) showing progressive normal cortical maturation in 13 subjects scanned biennially from ages 4 to 21 years. Blue-green scale colors correspond to lower GM densities, red-yellow to higher densities. Early regional maturation is a pattern of decreasing GM density involving primary and polar neocortical and limbic (not shown) areas. Later GM maturation occurs in the premotor, modality-specific association, and polymodal neocortex.

age range, shifts in activation were observed from frontal cortical regions to more posterior-inferior regions typical of the adult (Schmithorst, Holland, & Plante, 2006). A complex model for developmental changes in fMRI activation in a passive story-listening task has been presented, and generally supports a developmental shift in the strengths of functional connections between cortical language regions to the left hemisphere between 5 and 18 years of age (Karunanayaka et al., 2007).

The complex ordered developmental events just described provide a rich substrate for variable contributions to brain reserve. The numbers of neurons generated and their distribution through migration to form the cortical mantle, proper route-finding and establishment of synaptic connections between neurons, myelination of established fiber tracts, reinforcement of functional connections during the development of language and pruning of synapses in adolescence represent a few potential sources of these variations. These developmental events may therefore represent critical variables explaining different individual levels of resistance to the effects of brain pathology acquired years later in life.

2.2. Education and Cultural Influences

Education here refers to formal educational experiences, and cultural influences everything else, including family and friends, community, media, occupation, recreational activities and religion. Potential "biologic" exposures that might affect brain development or function, e.g., diet, water and air, toxin and drug exposures, and effects of medical disease are not considered.

The association between higher educational achievement and decreased risk of dementia is well established in the literature (Katzman, 1993). However, the level of education that best captures this association is controversial and an important issue because education before the end of adolescence is received in a critical developmental period biologically distinct from the adult, as described previously.

Some studies find the education effect only for no-education or primary school levels (Letenneur et al., 1999), but others find effects (odds ratio of about 2 for dementia) when defining low education as less than 8 years (Stern et al., 1994). Mixed model regression of education years in persons with pathologically defined AD but without clinical dementia showed that higher education was positively associated with this discordance (Roe, Xiong, Miller, & Morris, 2007). Expression of dementia in the presence of significant AD neuropathology was less likely by a factor of approximately 0.85 per year of additional education, supporting a brain reserve effect of education. An arbitrary education threshold was not used to produce this finding.

Related effects of education are that once symptoms begin, affected patients with higher educational attainment have greater atrophy (Kidron et al., 1997), greater metabolic abnormalities on functional imaging (Liao et al., 2005; Perneczky et al., 2006; Scarmeas et al., 2004; Y. Stern, Alexander, Prohovnik, & Mayeux, 1992), and greater cognitive decline to dementia (Unverzagt, Hui, Farlow, Hall, & Hendrie, 1998). These findings suggest that education may reduce or delay the expression of cognitive symptoms through a contribution to brain reserve, but once symptoms are expressed this reserve benefit is lost because the neuropathologic component of the underlying disease is more advanced at symptom onset.

Studies have shown that higher occupational attainment, which is difficult to disentangle from such factors as baseline intelligence, education and social advantage (Plassman et al., 1995), is associated with higher cognitive performance in late life and a lower likelihood of dementia (Andel et al., 2005; Potter, Plassman, Helms, Foster, & Edwards, 2006; Smyth et al., 2004), but again with an accelerated decline after symptoms begin (Andel, Vigen, Mack, Clark, & Gatz, 2006). Participation in cognitively stimulating activities may be the common link between brain reserve, education, and occupation (Wilson et al., 2002).

Recent studies have investigated the relationship of physical activity to cognitive performance, based in part on the hypothesis that such activity may stimulate adult neurogenesis and memory in mice (Steiner, Wolf, & Kempermann, 2006; Van der Borght, Havekes, Bos, Eggen, & Van der Zee, 2007). Although the contribution of physical fitness to overall health is unquestionable, the studies investigating a link of physical activity to brain reserve have been mixed (Kramer & Erickson, 2007; Scherder, Eggermont, Sergeant, & Boersma, 2007). Recent human and animal studies have not supported a contribution of "noncognitive" physical activity to cognitive impairment when environmentally stimulating cognitive activities are accounted for (Cracchiolo et al., 2007; Rovio et al., 2007). More work needs to be done in this area to supplement the education and occupation findings.

2.3. Genetic and Epigenetic Contributions to Brain Reserve

Here genetic refers to classic relationships between genotype and expression of traits and epigenetic denotes post-translational regulation and modification (e.g., DNA methylation) processes which may also alter that expression. The field of play for genetic and epigenetic mechanisms to influence brain reserve is quite large, including neurogenesis, neuronal migration and apportioning to functional regions, axonal route-finding and synaptogenesis, myelination, and synaptic modifications underlying particular types of neuronal plasticity. This section is restricted to an illustrative example of how these mechanisms could contribute to brain reserve. Neural plasticity is discussed further on in this review.

Migration of neurons from the ventricular zone to form the cortical mantle and the subsequent formation of the neocortical layers is a complex process whose genetic basis is just beginning to be understood (Rakic, 2007). This process occurs from 6-8 weeks gestation to four weeks before term (Sidman & Rakic, 1982). The migration process is guided by a specialized developmental scaffold consisting of the radial glia (Noctor, Martinez-Cerdeno, & Kriegstein, 2007; Rakic, 1990). The numerous anatomically distinct regions in the adult cortex are produced by preordained targeted neuronal migration to these regions and by continued elaboration of specific inter- and intra-regional connections (Rakic, 2007; Sidman & Rakic, 1982). With maturation, myelination patterns help to further define the adult architectonic areas (Bailey & Von Bonin, 1951; Brazier & Petsche, 1978). Genetic and epigenetic mechanisms orchestrate this process in part through determining neuronal numbers via mitotic division (Chenn & Walsh, 2002), motility patterns through calcium channel expression (Komuro & Rakic, 1996, 1998), filamin-A and MEKK4 (mitogen-activated protein kinase kinase kinase 4) signaling (Sarkisian et al., 2006), adhesion preference and strength via alpha3beta1 and alpha(v) integrins (Anton, Kreidberg, & Rakic, 1999), and termination of migration associated with expression of radial glial protein SPARC-like1 (secreted protein acidic and rich in cysteine-like 1) (Gongidi et al., 2004).

This process can go awry in association with human mutations in filamin-A, reelin (RELN), DCX, and aristaless-related homeobox protein (ARX), among many others (Barkovich, Kuzniecky, Jackson, Guerrini, & Dobyns, 2005). The complexity of interaction is illustrated by mutational variation in ARX, which has roles in several genetic cortical malformation disorders ranging from very mild to severe (Friocourt, Poirier, Rakic, Parnavelas, & Chelly, 2006; Kato et al., 2004; Stromme et al., 2002), and an important role in migration, neuronal proliferation, and maturation. Given the complexity of these multiple interactions in normal cortical development, a portion of brain reserve may well originate in genetic variations in ARX and other genes in the absence of mutation, influencing numerous critical points during cortical development, e.g., generation of raw primordial neuron numbers, development of and signaling by radial glia, neuronal place encoding in the ventricular and germinal matrix zones, targeting of neurons to their assigned region and layer of cortex, and formation of connections within and between cortical columns.

Neuropathology of AD acquired many years later develops within this field of genetic and epigenetically determined normal brain developmental variation. The amyloid precursor protein family, directly associated with AD neuropathology, also plays a key role in cortical development, particularly amyloid precursor like protein 1 (APLP1) (Lorent et al., 1995). Adaptor proteins interacting with this family, e.g., FE65, are required for normal cortical development (Guenette et al., 2006). Several other known or candidate genes in AD have important neurodevelopmental roles, particularly the presenilins and components of the Wnt signaling pathway, e.g., beta catenin (Chenn & Walsh, 2003; Ertekin-Taner, 2007; Malaterre, Ramsay, & Mantamadiotis, 2007; Wines-Samuelson & Shen, 2005). Recently the GRB-associated binding protein 2 (GAB2) gene has been found to be associated with late-onset AD, perhaps through interaction with other AD-related genes (E. M. Reiman et al., 2007). Much further work is needed to specify whether and to what degree genetic and epigenetic variation in specific pathways of brain development contribute to brain reserve and modification of symptom expression, apart from any role in AD neuropathology itself.

2.4. Normal Brain Alterations with Aging

Normal brain aging can be seen as consisting of consolidation, ongoing learning and adaptation on one hand, and erosion of brain reserve by involution on the other (Stern, 2002). On balance in successful brain aging, overall cognitive function is preserved. Here the focus is on structural anatomic alterations in the brain with normal aging.

Although there are some studies that suggest a somewhat different pattern (Allen, Bruss, Brown, & Damasio, 2005; Liu et al., 2003), others have shown that after a period of relative stability from the early 20s, there is a slow reduction in brain volume with age beginning at approximately 50-55 years that accelerates after age 70 (Figure 3A) (Coffey, Saxton, Ratcliff, Bryan, & Lucke, 1999; Coffey et al., 1992; Courchesne et al., 2000; DeCarli et al., 2005; Jernigan & Gamst, 2005; Lemaitre et al., 2005; Lim, Zipursky, Watts, & Pfefferbaum, 1992; Nagata et al., 1987; Pfefferbaum et al., 1994; Pfefferbaum, Sullivan, Swan, & Carmelli, 2000; Resnick, Pham, Kraut, Zonderman, & Davatzikos, 2003; Scahill et al., 2003). Voxel-based morphometric methods applied to magnetic resonance images have shown that GM volume and density reductions are global but the greatest reductions involve the parietal and frontal lobes (Good et al., 2001; Smith, Chebrolu, Wekstein, Schmitt, & Markesbery, 2007). Overall brain volume loss rates are on the order of 0.1 – 0.2% per year. There is little evidence of a selective regional volume loss exceeding global loss in the medial temporal lobe, the site of the earliest AD neuropathology. (DeCarli et al., 1994; Good et al., 2001; Smith, Chebrolu, Wekstein, Schmitt, & Markesbery, 2007). In this respect normal aging appears to differ from MCI and AD. It has been suggested that certain subcortical structures (substantia innominata) important in AD also undergo neuronal loss with aging, and that these losses may be partially reversed with intervention, e.g., nerve growth factor, at least in primates (Smith, Roberts, Gage, & Tuszynski, 1999).

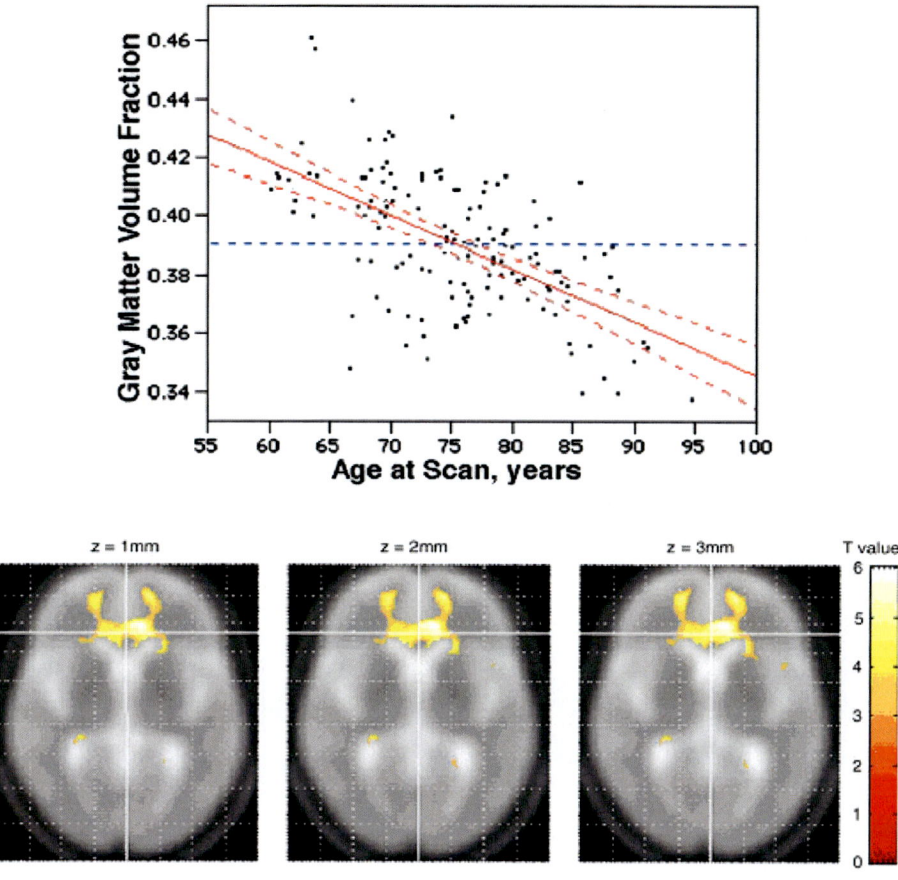

Figure 3. (A). Decline in TIV-normalized gray matter volume (fraction) with age in 122 normal subjects aged 58 – 95 years. In this cross-sectional study, there was a corresponding increase in cerebrospinal fluid volume (CSF) with age, but no change in global WM volume (Smith, Chebrolu, Wekstein, Schmitt, & Markesbery, 2007). (B). Localized WM volume decreases were demonstrated in the corpus callosum genu and frontal lobe WM (significant decreases in yellow on ascending axial slices with Talairach coordinates shown (Smith, Chebrolu, Wekstein, Schmitt, & Markesbery, 2007)).

The presence of global WM volume reductions with age remains an unresolved question, because some studies report it while others show no significant change (Good et al., 2001; Resnick et al., 2003; Smith, Chebrolu, Wekstein, Schmitt, & Markesbery, 2007; Walhovd et al., 2005). Other evidence suggest focal decreases in WM volume, specifically in the anterior corpus callosum and frontal lobe (Figure 3B) (Head, Snyder, Girton, Morris, & Buckner,

2005; Raz et al., 1997; Salat, Kaye, & Janowsky, 1999; Smith, Chebrolu, Wekstein, Schmitt, & Markesbery, 2007). However, the "frontal lobe hypothesis," used to explain declines in executive functions with age by positing a selective frontal involution, has been challenged (Greenwood, 2000).

The underlying morphologic correlates of brain volume reductions may include restricted neuronal losses, neuronal shrinkage, reduction in the size and extent of neuronal arbors and dendritic spines, and loss of myelin (Coleman & Flood, 1987; Dickstein et al., 2007; Flood & Coleman, 1988; Uylings & de Brabander, 2002). Therefore despite normal cognitive function, there may be diminished brain reserve measured generally by brain, and particularly cortical GM, volume loss in aging that may precede significant cognitive changes by many years. This cerebral atrophy is distinct from head circumference or total intracranial volume (TIV) as a measure of brain reserve (Schofield, Mosesson, Stern, & Mayeux, 1995; Wolf et al., 2004). Head circumference and TIV are better interpreted as a measure of in utero and postnatal brain developmental reserve (Bartholomeusz, Courchesne, & Karns, 2002).

2.5. Summary

Brain reserve is a useful general concept to help explain the ameliorating effects of education, occupation, and brain size on development of cognitive symptoms and dementia in late life. Specific potential components of this reserve discussed include the absolute number of neurons generated during early brain development, the targeted migration of these neurons to specific cortical regions, number and connections of cortical columns, editing and pruning of these connections during adolescence, and myelination of connecting fibers, most or all likely under genetic and epigenetic control. These components are consistent with known patterns of brain growth. Modifications of this substrate occur in the acquisition of language and under the influence of education, occupation, and other cognitive experiences that continue into adulthood.

Age-related decline in the morphologic characteristics and size or density of brain substance, particularly GM, is clearly established, and may diminish brain reserve in an age range when vulnerability to neurodegenerative pathologies is increased. The pattern of this age-related loss differs from the pattern seen in AD. Better understanding of the neurobiology of human brain development, even at its earliest stages, could provide important insights into how developmental processes leading to brain reserve could be enhanced. Currently little is known about specific contributions to the observed brain GM loss with normal aging, but there is no evidence that it is irreversible.

3. NEURONAL PLASTICITY

The concept that the brain undergoes continuous modification and change, particularly at neuronal synapses, dates back inevitably to Cajal, who explicitly used a Spanish term equivalent to "neuronal plasticity" (Cajal, 1995). However, the dynamic and muscular nature of synaptic change was demonstrated much more recently (Buonomano & Merzenich, 1998). The discussion of this vast topic is restricted to modification of synaptic connections in adult cortex as a response to two circumstances: (1) learning, broadly defined and (2) compensation for injuries to the nervous system.

3.1. Synaptic Changes with Learning

Evidence of synaptic plasticity is based on two types of observations: (1) changes in paradigmatic stimulation-induced evoked responses such as long-term potentiation or depression, and (2) morphologic changes in axons, boutons and dendritic spines (Deng & Dunaevsky, 2005; Lippman & Dunaevsky, 2005; Stettler, Yamahachi, Li, Denk, & Gilbert, 2006; Tsanov & Manahan-Vaughan, 2007). Alterations in synaptic signaling and structural proteins underlie many of these changes (Ethell & Pasquale, 2005; Gomes, Hampton, El-Sabeawy, Sabo, & McAllister, 2006; Sharma, Fong, & Craig, 2006; Tao-Cheng, 2006). That local synaptic responses and morphology undergo change with learning, e g, exposure to different sensory inputs, in the adult brain is no longer in doubt (Buonomano & Merzenich, 1998). What remains at issue is how these changes are related to larger scale reorganization of cortical representations or maps, and whether alterations in these maps and their interconnections form the essence of learning and learning-induced changes in behavior and cognition. For the present the assumption will be that this is the case and that in the adult human cortex, there is an ongoing normal process of orderly turnover, re-tuning and remodeling of synapses that underlies brain adaptations to a changing internal and external environment, constituting learning.

The questions raised here are whether biologic variations in the quality and efficiency of neuronal plasticity constitute an important component of brain reserve, and whether age-related losses in that quality might erode that reserve. In the absence of an accepted measure of the quality of synaptic plastic processes that would encompass variations within the biologically normal range, specific answers are not currently available. However, there are several identified biochemical pathways that have been associated with neuronal plasticity and learning where such biologic variations are likely to occur (Figure 4).

A signaling pathway with a nuclear target and positive and negative regulatory elements essential for the formation of long term memory, involving protein kinase A (PKA) and the cyclic AMP-response element binding protein (CREB), is well established (Abel & Kandel, 1998; Silva & Giese, 1994). Continued research has demonstrated a mitogen-activated protein kinase (MAPK) pathway that involves extracellular signal regulated kinases (erk) 1 and 2, which couple to protein and ion channel expression mechanisms, but that also modulate CREB through phosphorylation. A complementary pathway headed by Ras is inhibited by PKA; the two paths overlap by modulating MAPK kinase (MEK1 and 2) (Lonze & Ginty, 2002; Sweatt, 2001; Sweatt, Weeber, & Lombroso, 2003; Waltereit & Weller, 2003). The dual role of MAPK in memory and growth is incorporated within the concept that synaptic plasticity involves both changes in synaptic potential responses and in synaptic shape and size – physiological and morphological plasticity (Thomas & Huganir, 2004). Recent studies show that CREB-mediated transcription may be governed by an "off-on switch" provided by the phosphorylation state of eukaryotic translation initiation factor 2, subunit 1 alpha (eIF2alpha) (Costa-Mattioli et al., 2007).

Other molecular biology experiments suggest that presenilins and presenilin-interacting proteins (delta-catenin) play important roles in normal synaptic plasticity, in part through interactions with Notch and CREB signaling pathways (Israely et al., 2004; Marjaux, Hartmann, & De Strooper, 2004; Saura et al., 2004; Wines-Samuelson & Shen, 2005). Amyloid precursor protein (APP) is likewise associated with normal plasticity and is cleaved by gamma-secretase, which requires presenilin for this activity (Gralle & Ferreira, 2007; Israely et al., 2004; Mileusnic, Lancashire, & Rose, 2005; Schmidt et al., 2007; Turner, O'Connor, Tate, & Abraham, 2003; Zhang et al., 2007). Mutations in the genes for presenilin and APP cause autosomal dominant early onset Alzheimer's disease, but this does not exclude a potential contribution to brain reserve in their normal roles in synaptic plasticity (Marjaux et al., 2004; Mileusnic et al., 2005). There is clearly ample range in these pathways for variations in synaptic plasticity that could explain normal adult differences in learning and in variable erosion of brain reserve with age.

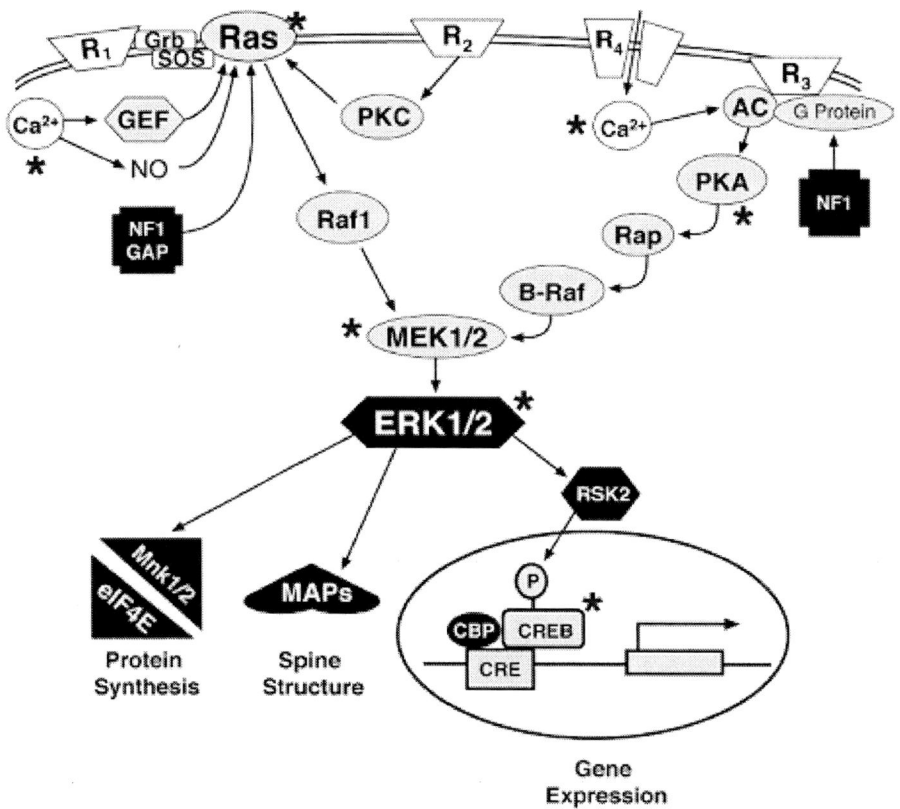

Figure 4. Diagram of a current model of neuronal learning modified (from Sweatt et al., 2003). Key participants in the signaling cascades discussed in the text are indicated by an asterisk (*) in the figure: Ras, Calcium (Ca2+), PKA, MEK1/2, ERF1/2, and CREB. End points of the cascades are gene transcription and spine structural and other protein synthesis. The letter R denotes a receptor, some of which are associated with coupling proteins. Further details can be found in the reference.

Neuronal plasticity is a "driven" process depending on signals from sensory inputs and from other connected brain regions. The quality of those inputs in part derives from the integrity of the peripheral sensory organs and receptors and the complexity and nature of environmental stimuli encountered. Because both of these factors may change with age, it is difficult to distinguish a normal adaptation of a functioning plastic mechanism due to changes in these factors and age-related changes in the plastic mechanisms themselves. Both could produce functional and even macroscopic morphologic changes seen with age because neural plasticity has both physiologic and morphologic components as described. There is little experimental evidence bearing on this distinction (Disterhoft & Oh, 2006; Drapeau, Montaron, Aguerre, & Abrous, 2007; Mothet et al., 2006). Some established changes with aging could affect the intrinsic mechanisms of neuronal plasticity, e.g., calcium dysregulation, because calcium is the key activator of both the Ras and MAPK signaling cascades (Foster, 2007; Mattson, 2007). However, despite the likely loss of brain reserve capacity, there is evidence that learning remains largely intact in normal successful aging and may be enhanced by intervention (Mahncke et al., 2006).

3.2. Compensations in Brain Injury

This section briefly explores what potentials and limits there might be in the capacity of plastic changes in the adult to compensate for an underlying neurodegenerative pathology. Certain forms of compensation such as modification of alternative macroscopic networks are not considered.

The dynamic nature of cortical maps has been shown by overlaps and shifts in receptive fields for the remaining innervated skin following limb amputation in young adults, demonstrating a potential capacity in this form of adult macroscopic cortical plasticity (Ramachandran, 2005a; Ramachandran & Rogers-Ramachandran, 2000). There is now a large literature in "applied neuronal plasticity," the attempt to move basic ideas concerning plasticity to the clinic to benefit brain injured patients to rehabilitate and compensate these injuries (Butefisch, 2006). Various evidence-based approaches have focused on motor rehabilitation during intensive physical therapy, including increasing cortical excitability in the injured region using repetitive transcranial magnetic stimulation (Butler & Wolf, 2007; Mally & Stone, 2007), increasing input to damaged cortex through peripheral sensory nerve stimulation (Castel-Lacanal, Gerdelat-Mas, Marque, Loubinoux, & Simonetta-Moreau, 2007; Conforto, Cohen, dos Santos, Scaff, & Marie, 2007), immobilization (constraint) of the unaffected extremity (Mark, Taub, & Morris, 2006; Sunderland & Tuke, 2005), and enhancement of synaptic activation using noradrenergic and dopaminergic drugs, e.g., amphetamine (Adkins & Jones, 2005; Ramic et al., 2006; Walker-Batson et al., 2001). Some of these efforts have been successful, but effects of many of these therapies have not been systematically demonstrated in randomized trials. Proof that improvements observed are in fact due to alterations in neuronal plasticity is not yet available. Nonetheless the principle that neural plasticity can usefully compensate for brain injury is supported by early studies in this field. Perhaps further research in this area will provide greater insight into why head injury is associated with AD (Blasko et al., 2004; Ikonomovic et al., 2004; Olsson et al., 2004; Szczygielski et al., 2005).

A surprising result from such studies is that brain "compensation" could impede recovery in the injured brain region through cortical inhibition from remaining intact regions, or by the expression of proteins that inhibit neural outgrowth and other aspects of plasticity, e.g., the Nogo-A protein (Chen et al., 2000). These results suggest that brain compensations for injury may be quite different from simple enhancement of the routine expression of neuronal plasticity, and that some compensations may actually reduce it. Deeper understanding of these compensatory mechanisms may lead to therapies designed to "unlock" normal plastic mechanisms to promote recovery in injured areas, e.g., antibodies to Nogo (Emerick & Kartje, 2004; Emerick, Neafsey, Schwab, & Kartje, 2003; Markus et al., 2005; Papadopoulos et al., 2002). There is a potential for such therapies in neurodegeneration if similar compensatory changes occur in those conditions, an unexplored possibility at present. Other compensatory mechanisms have been suggested to explain variability in Down's syndrome dementia, a dementia associated with AD neuropathology (Head, Lott, Patterson, Doran, & Haier, 2007).

3.3. Summary

Neural plasticity is the dominant known substrate for learning in the adult and has a detailed physiologic, morphologic, and molecular underpinning. Presenilin and APP have important normal roles in this substrate. Normal biologic variations in the quality and efficiency of plastic processes can plausibly explain differences in learning capacity between normal adults, a contribution to variations in brain reserve. Age-related decline in plasticity could be due to normal neural plastic responses to changes in sensory and environmental stimuli with age, age-related alterations in the character of plastic mechanisms themselves, or both. This decline appears small in relation to preservation of learning in aging, but it may signal the presence of an underlying erosion in brain reserve that can later be exposed by developing neuropathologic alterations in synapses. The brain compensates for injury at least in part through mechanisms of neural plasticity, but in some circumstances plasticity in the injured region may actually be reduced.

4. MILD COGNITIVE IMPAIRMENT (MCI) AND ALZHEIMER'S DISEASE (AD)

A simple listing of topic areas in this huge literature and its controversies would take many pages; therefore, simplification of complex issues is inevitable. Here the focus will be on key concepts related to MCI and AD that provide a basic foundation for the following section on presymptomatic detection.

4.1. Clinical Approach and Definition of MCI

Flicker et al showed progression of dementia in a group of longitudinally followed elderly subjects who initially had mild impairment on psychometric tests compared to controls (Flicker, Ferris, & Reisberg, 1991). Petersen et al emphasized and refined this concept of MCI and distinguished the amnestic form from other types (Petersen et al., 1999). The criteria of Petersen et al for amnestic MCI are most frequently used (Petersen, 1998): (a) memory complaints, preferably corroborated by an informant; (b) psychometrically defined memory impairment for age and education, but with preserved general cognitive function; (c) intact activities of daily living (ADLs); and (d) no clinical dementia. Subjects with amnestic MCI are by definition not demented, but are at greatly increased risk for the future development of dementia and in particular AD and vascular dementia, although rates of progression vary considerably across studies (Gauthier et al., 2006; Yaffe, Petersen, Lindquist, Kramer, & Miller, 2006).

The clinical approach is embodied in the clinical dementia rating (CDR), a formal structured interview of a patient and an informant by an experienced medical professional in which the opinion of degree of impairment is rated in five domains together with an overall rating score ranging from 0 to 3. In the proper context, this rating is highly reliable in the diagnosis and staging of MCI and dementia (Burke et al., 1988; Fillenbaum, Peterson, & Morris, 1996; Morris, 1997; Morris et al., 1991; Storandt, Grant, Miller, & Morris, 2006). A person who is asymptomatic by definition has a CDR of zero.

4.2. Neuropathology of MCI and AD

The diagnostic hallmarks of AD pathology are neuritic amyloid plaques, neuropil threads, and neurofibrillary tangles (Price et al., 1998). Neuronal and particularly synaptic losses accompany these hallmarks (Brun, Liu, & Erikson,

1995; P. Coleman, Federoff, & Kurlan, 2004; Masliah, Terry, DeTeresa, & Hansen, 1989; Scheff & Price, 2006; Scheff, Price, Schmitt, & Mufson, 2006; Small, 2004; Terry, 2006). The distribution of tangle neuropathology is incorporated into the Braak staging system for AD, which recognizes the earliest pathologic alterations in the transentorhinal cortex, followed by progressively greater tangle pathology in limbic regions, culminating in late stage widespread neocortical involvement (Braak, Alafuzoff, Arzberger, Kretzschmar, & Del Tredici, 2006; Braak, Braak, & Bohl, 1993). The rate of accumulation of this AD pathology is difficult to measure because in autopsy or biopsy only one time point can be sampled. Some studies have estimated a slightly increasing accumulation rate with age but the form of this acceleration is not clear, e.g., a power law or exponential rate increase (Dani, Pittella, Boehme, Hori, & Schneider, 1997; Miyasaka et al., 2005; Ohm, Muller, Braak, & Bohl, 1995).

In the presence of a clinical diagnosis of dementia and any acknowledged confounding brain pathologies, a neuropathologic diagnosis of AD can be made as low, intermediate, or high likelihood according to the National Institute on Aging and Reagan Institute criteria, which use semiquantitative estimates of CERAD neuritic plaque score and Braak staging. The stringency of these criteria depends on whether they are used for validation or other research, or for routine diagnostic purposes ("Consensus recommendations for the postmortem diagnosis of Alzheimer's disease. The National Institute on Aging, and Reagan Institute Working Group on Diagnostic Criteria for the Neuropathological Assessment of Alzheimer's Disease," 1997; Geddes et al., 1997; Jellinger, 1998; Jellinger & Bancher, 1998; Newell, Hyman, Growdon, & Hedley-Whyte, 1999; Wisniewski & Silverman, 1997). Neuropathology in amnestic MCI diagnosed in expert centers strongly resembles AD pathology in every respect except for lower levels of intensity and extent, supporting the construct of amnestic MCI as a symptomatic stage prior to clinically diagnosed AD (Bennett et al., 2006; Markesbery et al., 2006; Morris & Price, 2001; Petersen et al., 2006; Schmitt et al., 2000).

4.3. Non-AD Neuropathologies

The recognition of the contribution of non-AD neuropathologies to dementia, e.g., micro- and small infarcts and Lewy bodies, has complicated neuropathologic diagnosis (Kovari et al., 2007). Coexistence of these pathologies with AD pathology increases the likelihood of association with cognitive symptoms and clinical dementia diagnosis (Bennett, Schneider, Bienias, Evans, & Wilson, 2005;

Schneider, Arvanitakis, Bang, & Bennett, 2007; Snowdon et al., 1997). Because Lewy body pathology and AD pathology frequently coexist at autopsy, diagnosis of a Lewy body component distinct form AD remains a difficult clinical challenge (Metzler-Baddeley, 2007; Weisman et al., 2007).

An important theoretical issue is whether brain reserve should be considered in relation to one particular pathology or to all pathologies combined. If in relation to one pathology such as AD, remaining pathologies could be seen as eroding brain reserve relative to AD. However, simply because some pathologies, e.g., small infarcts, are common in aging does not mean they should be swept into the concept of reserve. This point is emphasized if one considers also including asymptomatic Lewy body pathology as eroding brain reserve relative to AD, which appears unwise because Lewy body pathology alone can cause dementia (Marui, Iseki, Kato, Akatsu, & Kosaka, 2004). Perhaps it is better is to consider brain reserve in relation to accumulating neuropathology whether individual or combined and assessing contributions from each type.

4.4. The Amyloid Hypothesis and Disease-Modifying Treatment

The amyloid hypothesis takes several forms, the gist of which is that abnormally increased production of particular beta-amyloid cleavage fragments of APP leads to a cascade of damaging effects on brain synapses and neurons (Walsh & Selkoe, 2004). The eventual consequence of this damage is MCI and ultimately, AD. The scientific support for this hypothesis would fill many volumes.

Several potential pharmacologic treatments for AD based on the amyloid hypothesis are in development, two of which have completed or will soon complete phase-III clinical trials seeking a Food and Drug Administration indication for disease modification. Tramiprosate binds beta amyloid and is thought to prevent formation of amyloid fibrils, and tarenflurbil is a selective amyloid beta 1-42 lowering agent (Gervais et al., 2007; Golde, 2006; Kukar et al., 2007). None of the medications approved beginning with tacrine in 1993 have had such an indication, so the introduction of these therapies is eagerly awaited by AD patients and their physicians. Neither drug has known effects on brain reserve or on non-AD pathologies.

4.5. Summary

Research has resulted in reliable clinical criteria for the diagnosis of amnestic MCI, which conceptually represents the earliest symptoms of mild memory loss presaging the onset of AD. Individuals with amnestic MCI have established neuropathology typical of AD, but the intensity and extent of that pathology is generally less than in AD, suggesting that further progression of pathology accompanies the further progression of mild memory symptoms to dementia. Disease modifying treatment is in clinical trial and is expected to slow the progression of pathology and therefore the rate of clinical progression from MCI to AD. The presence of other pathologies increases the likelihood that a given level of AD pathology will be associated with dementia. Disease modifying treatment would not decrease this association but could reduce the level of AD pathology and consequent dementia.

5. PRESYMPTOMATIC DETECTION OF MCI AND AD

The purpose of this section is to discuss evidence that AD pathology is present in normal persons for many years prior to the period of vulnerability for late-onset AD. Potential imaging, biofluid, clinical and psychometric methods for detecting effects of this pathology and predicting onset of early AD symptoms are presented.

5.1. Evidence for Presymptomatic Neuropathology in AD

Neurofibrillary pathology characteristic of AD but restricted to entorhinal and hippocampal areas (Braak stages I & II) was present in over 50% of autopsies performed between ages 50 and 60 years as part of a larger study of 2,661 staged neuropathologic examinations in the age range 25 – 95 years (Braak & Braak, 1998). The age range of 50-60 is below the usual expected onset of clinical late-onset AD (McKhann et al., 1984). The presence of AD pathology in older, carefully evaluated nondemented subjects has now been documented in several subsequent studies (Bennett et al., 2006; Davis, Schmitt, Wekstein, & Markesbery, 1999; Driscoll et al., 2006; Markesbery et al., 2006; Schmitt et al., 2000). Two key issues raised by these findings are: (1) whether this pathology is related to subtle cognitive decline to levels that remain within the normal range and that are insufficient to produce functional impairments, and (2) the characteristics of AD pathologic progression demarking the transition to the symptomatic state, e.g., MCI. The issue of cognitive change is discussed later in this section; here the neuropathologic correlates of transition to MCI are briefly discussed. .

Recent studies have shown that likelihood of early memory and related symptoms is related to two basic factors: the presence of coexistent non-AD pathologies, e.g., microinfarcts, and the extent or stage of AD pathology (Bennett et al., 2006; Bennett et al., 2005; Petersen et al., 2006; Schmitt et al., 2000; Schneider et al., 2007). Increased numbers of limbic neurofibrillary tangles and the presence of neocortical neuritic plaque pathology appear to best distinguish persons meeting clinical criteria for MCI from normal (Markesbery et al., 2006; Morris et al., 1991; Morris & Price, 2001). Nonetheless similar mild to moderate levels of AD pathology seen in MCI are also observed in well characterized subjects with normal levels of cognition and functional performance. The concept of brain reserve and its several components as variables explaining this discrepancy have been discussed in a previous section.

5.2. Imaging Biomarkers

Development of imaging as a technique to assess the brain in normal subjects prior to MCI and dementia may have different goals that should be distinguished: (1) detecting the presence of morphologic and functional changes due to the presence of underlying asymptomatic AD and other pathologies, (2) characterizing brain features related to brain reserve and changes in that reserve, and (3) mapping brain compensatory changes related to advancing neuropathology or to involutional age-related erosion of brain reserve. The potential for confounding these different goals depending on limitations of imaging techniques or of practical experimental designs should be kept in mind.

Much imaging research has been based on the hypothesis that because the earliest pathologic alterations in AD involve the entorhinal cortex and other medial temporal limbic structures, these regions will be the most salient in predicting normal subjects who would later develop MCI or AD (Kaye et al., 1997). Absolute entorhinal and hippocampal volumes are decreased and rates of volume change are increased in MCI (Convit et al., 2000; Jack et al., 1999; Jack et al., 2004; Xu et al., 2000). However, some studies have shown that general measures of brain atrophy rate are as salient as measures of entorhinal and hippocampal atrophy rate in predicting conversion from normal to MCI (Jack et al., 2004; Jack et al., 2005) or cognitive decline without dementia (Adak et al.,

2004), suggesting that there may be a separate brain reserve component outside the medial temporal target region of early AD neuropathology (Chetelat et al., 2005; Convit et al., 2000; Jack et al., 2000; Karas et al., 2004; Mueller et al., 1998; Pennanen et al., 2005). Others have demonstrated decreased GM density in cognitively normal apolipoprotein (APOE) E4 allele carriers at risk for AD (Wishart, Saykin, McAllister et al., 2006).

A recent study demonstrated that medial temporal and left parietal GM volumes in normal subjects predicted MCI within five years with 76% accuracy that was further enhanced to 87% by combining GM volume with a cognitive measure, raw Wechsler Memory Scale score (Figure 5A) (Smith, Chebrolu, Wekstein, Schmitt, Jicha et al., 2007). Lower resting glucose metabolism using positron emission tomography (PET) in the left angular, mid temporal, and mid frontal gyri predict decline on a global cognitive measure, the Mini Mental State Examination, and decreased medial temporal volumes predict decline on delayed memory in normal subjects at baseline followed subsequently over 3.8 years (Jagust et al., 2006). Ten percent of these subjects developed MCI or dementia, but the correlation remained significant without them, suggesting these imaging measures may predict MCI or dementia in normal subjects.

Functional imaging studies have shown alterations in brain activation in subjects at increased risk of AD years before the onset of cognitive symptoms. Decreased PET resting glucose metabolism in the posterior cingulate and parietal regions has been demonstrated in normal subjects at increased risk of early-onset AD due presenilin and APP mutations, and in subjects with a family history of late-onset AD and at least one apolipoprotein (APOE) E4 allele (Johnson et al., 2001; Reiman et al., 1996; G. W. Small et al., 2000; Stefanova et al., 2002). Amyloid imaging, a PET-based technique for quantitating the burden of brain amyloid, is conceptually appealing and holds promise for presymptomatic detection (Lockhart et al., 2007; Nordberg, 2007; Rowe et al., 2007). The role of this new technique for presymptomatic detection is not yet established.

Active functional MRI and PET results have not been entirely consistent. Some studies have demonstrated decreased activation in prefrontal activity in normal high-risk (APOE4 carrier) subjects during working memory tasks, in ventral temporal regions during picture naming, and in medial temporal regions during episodic memory encoding (Figure 5B) (Elgh, Larsson, Eriksson, & Nyberg, 2003; C. D. Smith et al., 1999; C. D. Smith et al., 2005; Trivedi et al., 2006). Other studies have demonstrated increased parietal activation associated with verbal fluency tasks and increased medial temporal activation during episodic memory encoding or retrieval (Figure 5C) (Bondi, Houston, Eyler, & Brown, 2005; Bookheimer et al., 2000; Fleisher et al., 2005; Smith et al., 2002;

Sperling, 2007; Wishart, Saykin, Rabin et al., 2006). These functional studies differ in the age of subjects, inclusion of family history with the APOE4 carrier state, nature of tasks performed during imaging, thoroughness of cognitive testing, and methods of analyses. Nonetheless, these studies suggest that active functional imaging holds promise as class of imaging biomarkers for presymptomatic brain alterations in normal persons at risk for AD.

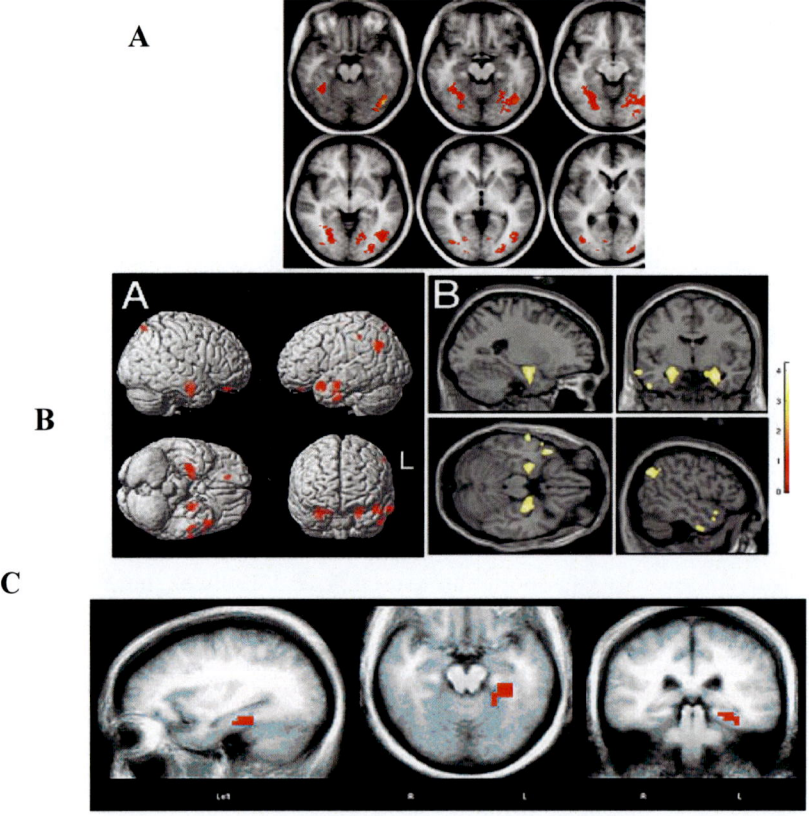

Fig. 5. Illustration of structural and functional imaging findings in asymptomatic normal subjects at risk of AD. (A) Regions of significant volume decrease in the anteromedial temporal lobes and left angular gyrus region in 23 normal subjects destined to develop MCI vs. 113 normal subjects who remained normal over 5 years follow-up assessment (Smith, Chebrolu, Wekstein, Schmitt, Jicha et al., 2007). (B) Decreased ventral temporal activation during picture naming in 14 presymptomatic normal subjects at increased risk of AD vs. 12 low-risk subjects (Smith et al., 1999). (C) Left medial temporal activation increase during paired word associates learning in 10 presymptomatic normal subjects at increased risk of AD vs. 10 low-risk subjects (Fleisher et al., 2005). In (A) & (B) the red and yellow overlay indicates a decrease in the comparison; in (C) red indicates an increase.

5.3. Serum and CSF Biomarkers

The ratio between beta amyloid isoforms 1-42 and 1-40 in plasma and CSF has been shown to associate with or predict risk of later MCI or AD in normal subjects in separate studies (Fagan et al., 2007; Graff-Radford et al., 2007). However, the relationship between plasma and CSF amyloid has not yet been fully clarified (Freeman, Raju, Hyman, Frosch, & Irizarry, 2007). A combination of tau, phospho-tau, and beta amyloid 1-42 CSF levels may improve prediction of conversion from MCI to AD. Other approaches to improve sensitivity and specificity of CSF analytes have included measurement of specific oxidation products, e.g. 4-hydroxynonenal, and protein patterning to detect a profile of neuropathology-related molecular alterations (Finehout, Franck, Choe, Relkin, & Lee, 2007; Lovell, Xie, & Markesbery, 1998). Studies have not yet examined these or similar promising CSF biomarker approaches to predict later cognitive symptoms of MCI in normal subjects.

5.4. Psychometric Assessments to Predict MCI or AD

That cognitive performance in early adulthood may predict performance in late life would not be surprising. For example, performance on a military examination, the Army General Classification Test accounted for 21% of the variance in cognitive performance measured 50 years later and education accounted for 16.7% (Plassman et al., 1995). However, cognitive performance in childhood and early adulthood has also been linked to later expression of dementia symptoms. Performance of 11-year-olds on a standardized mental ability test in Scotland in 1932 was lower in survivors who developed dementia after age 65 years than in those who did not. In a classic longitudinal study of nuns, idea density and grammatical complexity judged from essays written when subjects entered their religious order at age 22 was lower in those who developed dementia many years later (Snowdon et al., 1996). Investigators in the Framingham study found that derived memory scores were lower in normal subjects who developed dementia an average of 22 years later (Elias et al., 2000). These early detected differences in performance and ability are most likely determined by developmental differences discussed previously under the heading of brain reserve.

The literature on presymptomatic prediction is confounded by newly improved concepts and clinical techniques for the detection of MCI, which themselves include psychometric assessment as one of the criteria for this diagnosis. This moving target makes it more difficult to interpret literature demonstrating presymptomatic cognitive declines, particularly on subtle memory measures (Blacker et al., 2007; Crystal et al., 1996; Grober, Lipton, Hall, & Crystal, 2000; Masur, Sliwinski, Lipton, Blau, & Crystal, 1994; Small, Stern, Tang, & Mayeux, 1999; Tierney, Yao, Kiss, & McDowell, 2005). There is no doubt that decline on psychometric measures occurs prior to dementia.

The clinical and neuropsychological approaches to presymptomatic detection encompass a range: the CDR is a structured clinical evaluation that does not include formal psychometric testing (Morris, 1997); the diagnosis of MCI may include a CDR-like clinical assessment but also explicitly incorporates psychometric testing results within its criteria (Kelley & Petersen, 2007), and finally, many studies used cognitive assessments alone to determine pre-dementia states (Blacker et al., 2007). The types of evaluations bracketing the two ends of this range are fundamentally different and each approach has advantages and defects. The CDR is an individual determination that does not require adjustments for age or education (norms), but it does require clinical expertise, judgment, and time (Morris, 1997). The recently developed AD8 may overcome some of this limitation (Galvin, Roe, Xiong, & Morris, 2006). Psychometric evaluations provide quantitative results in detailed cognitive domains and are easier to implement in large-scale studies but scores must be adjusted for several confounds, including age, and may be subject to greater individual variation over time (Grober et al., 2000; Small et al., 1999; Tierney et al., 2005).

MCI diagnosis is the middle way, incorporating advantages of both approaches. However, in this case there is more data to consider, which may lead to the problem of reconciling the two when results conflict. One group that has used both methods independently reports no psychometrically-defined cognitive impairment or decline in longitudinally followed subjects prior to clinical detection, even in the presence of AD pathology (Goldman et al., 2001; Storandt et al., 2006). Others have drawn similar conclusions (Driscoll et al., 2006). This suggests that other methods are needed to detect AD pathology before MCI (Morris & Price, 2001).

5.5. Summary

AD neuropathology and often other pathologies are already present when the earliest memory and related cognitive symptoms first appear. In many cases these symptoms herald a progressive clinical dementia. These observations imply that neuropathology was present for some interval when the affected individual was cognitively and behaviorally indistinguishable from other normal persons without AD pathology. Accumulating data suggest this interval can be measured in decades. Current evidence indicates that the complementary clinical and psychometric detection methods available today may not make this critical distinction. If disease modifying treatments soon to be available slow the progress of AD pathology, it follows that these treatments could delay the onset of symptoms or perhaps prevent their appearance entirely. Imaging biomarkers and serum and CSF analytes may provide the means to detect presymptomatic pathology and follow its evolution to measure treatment effects in future preventive clinical trials.

6. OVERALL SUMMARY AND CONCLUSIONS

6.1. The Argument for Presymptomatic Detection

Here the use of qualifiers, e.g., "might", and "may", is dropped to emphasize the fundamental argument. Presymptomatic detection refers specifically to detection of late-onset AD.

Intrinsic or acquired brain properties confer variable resistance to the effects of neuropathology, modifying whether and when that neuropathology is expressed as symptoms, changes in cognitive performance, or behavior. These properties include developmental factors such as original neuronal number and the extent of synaptic and fiber pruning during adolescence. Rates of formation and maintenance of synaptic connections and of myelination of nerve fibers are critical sources of normal variation both in development and the adult. Educational and other sources of cognitive stimulation also contribute to brain reserve by increasing the redundancy or multiplicity of brain associative connections formed during childhood and throughout adulthood. Genetic and epigenetic mechanisms underlie these developmental events and the synaptic and neuronal molecular mechanisms of learning in the adult, constituting a source of normal biologic variation in reserve. Genes for APP and presenilin known to cause early-onset AD when mutated are essential to this variation. As the brain ages, there is a slow loss of global GM volume and a focal loss of frontal WM, eroding this reserve. Normal intrinsic neuronal plasticity compensates for a portion of this erosion and, in successful aging, cognition is preserved into late life.

AD neuropathology is present in significant amounts by at least age 50 years, and increases in intensity and extent despite preservation of cognition at this age.

Imaging evidence demonstrates presymptomatic alterations in brain function in this age range in persons at increased risk of AD, representing brain compensation for the presence of AD pathology. Typically after age 65, whether and when cognitive symptoms appear in persons who have a given level of pathology is determined by: (1) the variable level of reserve set by its several underlying mechanisms or (2) exhaustion of existing compensation, or both. The AD neuropathologic process itself may accelerate at this threshold. In this view, the appearance of symptoms marks a critical point where the future clinical trajectory is determined by AD and other pathologies; intrinsic reserve and compensation can no longer have significant effects. Even in the ideal circumstance where AD pathology is halted completely by specific disease-modifying treatment, other accumulating pathologies and age-related loss of brain substance can cause worsening.

The situation with disease-modification in the presymptomatic period is different. Slowing of the AD pathologic process before symptoms appear delays their appearance even if mild, allows compensatory mechanisms to address a slower progression of pathology, and saves the reserve needed to offset the effects of other pathologies, e.g., microinfarcts. If neuropathology accelerates with age or onset of symptoms, early intervention has compound effects.

6.2. Future Directions

The program suggested to address these conclusions is to focus more attention on presymptomatic detection based on the promising early results in this field. We need to know more about human developmental events and their molecular underpinning from the perspective of their potential effect in later life. In particular, the role of genes associated with early and late-onset AD, e.g., APP, presenilin, SORL1 and GAB2 in development should be further explored. We need to know more about the molecular events involved in normal adult synaptic plasticity and in related compensatory mechanisms, e.g., No-goA and the AD-associated genes. Further longitudinal imaging and clinical study of normal aging with continued follow-up into the MCI transition is needed to discover alterations prior to symptoms that are invisible to present clinical methods.

Excellent and laudable efforts are ongoing in many of these areas, but an organized focus from the perspective of late-life dementia is not always evident. Understandably much previous work has been devoted to characterizing symptomatic AD states, but some of the many brilliant minds behind these investigations could also advance the cause of presymptomatic detection for the reasons given here.

ACKNOWLEDGEMENTS

Apologies in advance for any omissions of research contributions not cited here; there is no intent to neglect or inaccurately represent anyone's work. Gratitude is owed to the many brilliant investigators who have made important contributions to AD research so that patients and families may benefit. The author thanks Dr. W.R. Markesbery for his mentorship, and the NIH for support through NIH grants P30 AG028383 and R01 NS36660.

REFERENCES

Abel, T., & Kandel, E. (1998). Positive and negative regulatory mechanisms that mediate long-term memory storage. *Brain Res Brain Res Rev*, 26(2-3), 360-378.

Adak, S., Illouz, K., Gorman, W., Tandon, R., Zimmerman, E. A., Guariglia, R., Moore, M. M., & Kaye, J. A. (2004). Predicting the rate of cognitive decline in aging and early Alzheimer disease. *Neurology*, 63(1), 108-114.

Adkins, D. L., & Jones, T. A. (2005). D-amphetamine enhances skilled reaching after ischemic cortical lesions in rats. *Neurosci Lett*, 380(3), 214-218.

Allen, J. S., Bruss, J., Brown, C. K., & Damasio, H. (2005). Normal neuroanatomical variation due to age: the major lobes and a parcellation of the temporal region. *Neurobiol Aging*, 26(9), 1245-1260; discussion 1279-1282.

Andel, R., Crowe, M., Pedersen, N. L., Mortimer, J., Crimmins, E., Johansson, B., & Gatz, M. (2005). Complexity of work and risk of Alzheimer's disease: a population-based study of Swedish twins. *J Gerontol B Psychol Sci Soc Sci*, 60(5), P251-258.

Andel, R., Vigen, C., Mack, W. J., Clark, L. J., & Gatz, M. (2006). The effect of education and occupational complexity on rate of cognitive decline in Alzheimer's patients. *J Int Neuropsychol Soc*, 12(1), 147-152.

Anton, E. S., Kreidberg, J. A., & Rakic, P. (1999). Distinct functions of alpha3 and alpha(v) integrin receptors in neuronal migration and laminar organization of the cerebral cortex. *Neuron*, 22(2), 277-289.

Arnold, S. E., & Trojanowski, J. Q. (1996). Human fetal hippocampal development: I. Cytoarchitecture, myeloarchitecture, and neuronal morphologic features. *J Comp Neurol*, 367(2), 274-292.

Bailey, P., & Von Bonin, G. (1951). *The isocortex of man*. Urban: University of Illinois Press.

Barkovich A.J., Kjos, B. O., Jackson, D. E., & Norman, D. (1988). Normal maturation of the neonatal and infant brain: MR imaging at 1.5 T. *Radiology* 166, 173-180.

Barkovich, A. J., Kuzniecky, R. I., Jackson, G. D., Guerrini, R., & Dobyns, W. B. (2005). A developmental and genetic classification for malformations of cortical development. *Neurology*, 65(12), 1873-1887.

Bartholomeusz, H. H., Courchesne, E., & Karns, C. M. (2002). Relationship between head circumference and brain volume in healthy normal toddlers, children, and adults. *Neuropediatrics*, 33(5), 239-241.

Benes, F. M., Turtle, M., Khan, Y., & Farol, P. (1994). Myelination of a key relay zone in the hippocampal formation occurs in the human brain during childhood, adolescence, and adulthood. *Arch Gen Psychiatry*, 51(6), 477-484.

Bennett, D. A., Schneider, J. A., Arvanitakis, Z., Kelly, J. F., Aggarwal, N. T., Shah, R. C., & Wilson, R. S. (2006). Neuropathology of older persons without cognitive impairment from two community-based studies. *Neurology*, 66(12), 1837-1844.

Bennett, D. A., Schneider, J. A., Bienias, J. L., Evans, D. A., & Wilson, R. S. (2005). Mild cognitive impairment is related to Alzheimer disease pathology and cerebral infarctions. *Neurology*, 64(5), 834-841.

Blacker, D., Lee, H., Muzikansky, A., Martin, E. C., Tanzi, R., McArdle, J. J., Moss, M., & Albert, M. (2007). Neuropsychological measures in normal individuals that predict subsequent cognitive decline. *Arch Neurol*, 64(6), 862-871.

Blasko, I., Beer, R., Bigl, M., Apelt, J., Franz, G., Rudzki, D., Ransmayr, G., Kampfl, A., & Schliebs, R. (2004). Experimental traumatic brain injury in rats stimulates the expression, production and activity of Alzheimer's disease beta-secretase (BACE-1). *J Neural Transm*, 111(4), 523-536.

Bondi, M. W., Houston, W. S., Eyler, L. T., & Brown, G. G. (2005). fMRI evidence of compensatory mechanisms in older adults at genetic risk for Alzheimer disease. *Neurology*, 64(3), 501-508.

Bookheimer, S. Y., Strojwas, M. H., Cohen, M. S., Saunders, A. M., Pericak-Vance, M. A., Mazziotta, J. C., & Small, G. W. (2000). Patterns of brain activation in people at risk for Alzheimer's disease. *N Engl J Med*, 343(7), 450-456.

Borenstein Graves, A., Mortimer, J. A., Bowen, J. D., McCormick, W. C., McCurry, S. M., Schellenberg, G. D., & Larson, E. B. (2001). Head circumference and incident Alzheimer's disease: modification by apolipoprotein E. *Neurology*, 57(8), 1453-1460.

Braak, H., Alafuzoff, I., Arzberger, T., Kretzschmar, H., & Del Tredici, K. (2006). Staging of Alzheimer disease-associated neurofibrillary pathology using paraffin sections and immunocytochemistry. *Acta Neuropathol (Berl)*, 112(4), 389-404.

Braak, H., & Braak, E. (1998). Alzheimer's disease starts in early adulthood. *Neurobiology of Aging*, 19, S78.

Braak, H., Braak, E., & Bohl, J. (1993). Staging of Alzheimer-related cortical destruction. *Eur Neurol*, 33(6), 403-408.

Brazier, M. A. B., & Petsche, H. (Eds.). (1978). *Architectonics of the cerebral cortex.* New York: Raven Press.

Brun, A., Liu, X., & Erikson, C. (1995). Synapse loss and gliosis in the molecular layer of the cerebral cortex in Alzheimer's disease and in frontal lobe degeneration. *Neurodegeneration*, 4(2), 171-177.

Buonomano, D. V., & Merzenich, M. M. (1998). Cortical plasticity: from synapses to maps. *Annu Rev Neurosci*, 21, 149-186.

Burke, W. J., Miller, J. P., Rubin, E. H., Morris, J. C., Coben, L. A., Duchek, J., Wittels, I. G., & Berg, L. (1988). Reliability of the Washington University Clinical Dementia Rating. *Arch Neurol*, 45(1), 31-32.

Butefisch, C. M. (2006). Neurobiological bases of rehabilitation. *Neurol Sci*, 27 Suppl 1, S18-23.

Butler, A. J., & Wolf, S. L. (2007). Putting the brain on the map: use of transcranial magnetic stimulation to assess and induce cortical plasticity of upper-extremity movement. *Phys Ther*, 87(6), 719-736.

Castel-Lacanal, E., Gerdelat-Mas, A., Marque, P., Loubinoux, I., & Simonetta-Moreau, M. (2007). Induction of cortical plastic changes in wrist muscles by paired associative stimulation in healthy subjects and post-stroke patients. *Exp Brain Res*, 180(1), 113-122.

Chen, M. S., Huber, A. B., van der Haar, M. E., Frank, M., Schnell, L., Spillmann, A. A., Christ, F., & Schwab, M. E. (2000). Nogo-A is a myelin-associated neurite outgrowth inhibitor and an antigen for monoclonal antibody IN-1. *Nature*, 403(6768), 434-439.

Chenn, A., & Walsh, C. A. (2002). Regulation of cerebral cortical size by control of cell cycle exit in neural precursors. *Science*, 297(5580), 365-369.

Chenn, A., & Walsh, C. A. (2003). Increased neuronal production, enlarged forebrains and cytoarchitectural distortions in beta-catenin overexpressing transgenic mice. *Cereb Cortex*, 13(6), 599-606.

Chetelat, G., Landeau, B., Eustache, F., Mezenge, F., Viader, F., de la Sayette, V., Desgranges, B., & Baron, J. C. (2005). Using voxel-based morphometry to map the structural changes associated with rapid conversion in MCI: a longitudinal MRI study. *Neuroimage*, 27(4), 934-946.

Coffey, C. E., Saxton, J. A., Ratcliff, G., Bryan, R. N., & Lucke, J. F. (1999). Relation of education to brain size in normal aging: implications for the reserve hypothesis. *Neurology*, 53(1), 189-196.

Coffey, C. E., Wilkinson, W. E., Parashos, I. A., Soady, S. A., Sullivan, R. J., Patterson, L. J., Figiel, G. S., Webb, M. C., Spritzer, C. E., & Djang, W. T. (1992). Quantitative cerebral anatomy of the aging human brain: a cross-sectional study using magnetic resonance imaging. *Neurology*, 42(3 Pt 1), 527-536.

Coleman, P., Federoff, H., & Kurlan, R. (2004). A focus on the synapse for neuroprotection in Alzheimer disease and other dementias. *Neurology*, 63(7), 1155-1162.

Coleman, P. D., & Flood, D. G. (1987). Neuron numbers and dendritic extent in normal aging and Alzheimer's disease. *Neurobiol Aging*, 8(6), 521-545.

Conforto, A. B., Cohen, L. G., dos Santos, R. L., Scaff, M., & Marie, S. K. (2007). Effects of somatosensory stimulation on motor function in chronic cortico-subcortical strokes. *J Neurol*, 254(3), 333-339.

Consensus recommendations for the postmortem diagnosis of Alzheimer's disease. The National Institute on Aging, and Reagan Institute Working Group on Diagnostic Criteria for the Neuropathological Assessment of Alzheimer's Disease. (1997). *Neurobiol Aging*, 18(4 Suppl), S1-2.

Convit, A., de Asis, J., de Leon, M. J., Tarshish, C. Y., De Santi, S., & Rusinek, H. (2000). Atrophy of the medial occipitotemporal, inferior, and middle temporal gyri in non-demented elderly predict decline to Alzheimer's disease. *Neurobiol Aging*, 21(1), 19-26.

Costa-Mattioli, M., Gobert, D., Stern, E., Gamache, K., Colina, R., Cuello, C., Sossin, W., Kaufman, R., Pelletier, J., Rosenblum, K., Krnjevic, K., Lacaille, J. C., Nader, K., & Sonenberg, N. (2007). eIF2alpha phosphorylation bidirectionally regulates the switch from short- to long-term synaptic plasticity and memory. *Cell*, 129(1), 195-206.

Courchesne, E., Chisum, H. J., Townsend, J., Cowles, A., Covington, J., Egaas, B., Harwood, M., Hinds, S., & Press, G. A. (2000). Normal brain development and aging: quantitative analysis at in vivo MR imaging in healthy volunteers. *Radiology*, 216(3), 672-682.

Cracchiolo, J. R., Mori, T., Nazian, S. J., Tan, J., Potter, H., & Arendash, G. W. (2007). Enhanced cognitive activity-over and above social or physical activity-is required to protect Alzheimer's mice against cognitive impairment, reduce Abeta deposition, and increase synaptic immunoreactivity. *Neurobiol Learn Mem*.

Crystal, H., Dickson, D., Sliwinski, M., Masur, D., Blau, A., & Lipton, R. B. (1996). Associations of status and change measures of neuropsychological function with pathologic changes in elderly, originally nondemented subjects. *Arch Neurol*, 53, 82-87.

Dani, S. U., Pittella, J. E., Boehme, A., Hori, A., & Schneider, B. (1997). Progressive formation of neuritic plaques and neurofibrillary tangles is exponentially related to age and neuronal size. A morphometric study of three geographically distinct series of aging people. *Dement Geriatr Cogn Disord*, 8(4), 217-227.

Davis, D. G., Schmitt, F. A., Wekstein, D. R., & Markesbery, W. R. (1999). Alzheimer neuropathologic alterations in aged cognitively normal subjects. *J Neuropathol Exp Neurol*, 58(4), 376-388.

DeCarli, C., Massaro, J., Harvey, D., Hald, J., Tullberg, M., Au, R., Beiser, A., D'Agostino, R., & Wolf, P. A. (2005). Measures of brain morphology and infarction in the framingham heart study: establishing what is normal. *Neurobiol Aging*, 26(4), 491-510.

DeCarli, C., Murphy, D. G., Gillette, J. A., Haxby, J. V., Teichberg, D., Schapiro, M. B., & Horwitz, B. (1994). Lack of age-related differences in temporal lobe volume of very healthy adults. *AJNR Am.J Neuroradiol.*, 15, 689-696.

Deng, J., & Dunaevsky, A. (2005). Dynamics of dendritic spines and their afferent terminals: spines are more motile than presynaptic boutons. *Dev Bioll*, 277(2), 366-377.

Dickstein, D. L., Kabaso, D., Rocher, A. B., Luebke, J. I., Wearne, S. L., & Hof, P. R. (2007). Changes in the structural complexity of the aged brain. *Aging Cell*, 6(3), 275-284.

Disterhoft, J. F., & Oh, M. M. (2006). Learning, aging and intrinsic neuronal plasticity. *Trends Neurosci*, 29(10), 587-599.

Drapeau, E., Montaron, M. F., Aguerre, S., & Abrous, D. N. (2007). Learning-induced survival of new neurons depends on the cognitive status of aged rats. *J Neurosci*, 27(22), 6037-6044.

Driscoll, I., Resnick, S. M., Troncoso, J. C., An, Y., O'Brien, R., & Zonderman, A. B. (2006). Impact of Alzheimer's pathology on cognitive trajectories in nondemented elderly. *Ann Neurol*, 60(6), 688-695.

Elgh, E., Larsson, A., Eriksson, S., & Nyberg, L. (2003). Altered prefrontal brain activity in persons at risk for Alzheimer's disease: an fMRI study. *Int Psychogeriatr*, 15(2), 121-133.

Elias, M. F., Beiser, A., Wolf, P. A., Au, R., White, R. F., & D'Agostino, R. B. (2000). The preclinical phase of alzheimer disease: A 22-year prospective study of the Framingham Cohort. *Arch Neurol*, 57(6), 808-813.

Emerick, A. J., & Kartje, G. L. (2004). Behavioral recovery and anatomical plasticity in adult rats after cortical lesion and treatment with monoclonal antibody IN-1. *Behav Brain Res*, 152(2), 315-325.

Emerick, A. J., Neafsey, E. J., Schwab, M. E., & Kartje, G. L. (2003). Functional reorganization of the motor cortex in adult rats after cortical lesion and treatment with monoclonal antibody IN-1. *J Neurosci*, 23(12), 4826-4830.

Ertekin-Taner, N. (2007). Genetics of Alzheimer's disease: a centennial review. *Neurol Clin*, 25(3), 611-667.

Ethell, I. M., & Pasquale, E. B. (2005). Molecular mechanisms of dendritic spine development and remodeling. *Prog Neurobiol*, 75(3), 161-205.

Fagan, A. M., Roe, C. M., Xiong, C., Mintun, M. A., Morris, J. C., & Holtzman, D. M. (2007). Cerebrospinal fluid tau/beta-amyloid(42) ratio as a prediction of cognitive decline in nondemented older adults. *Arch Neurol*, 64(3), 343-349.

Fillenbaum, G. G., Peterson, B., & Morris, J. C. (1996). Estimating the validity of the clinical Dementia Rating Scale: the CERAD experience. Consortium to Establish a Registry for Alzheimer's Disease. *Aging (Milano)*, 8(6), 379-385.

Finehout, E. J., Franck, Z., Choe, L. H., Relkin, N., & Lee, K. H. (2007). Cerebrospinal fluid proteomic biomarkers for Alzheimer's disease. *Ann Neurol*, 61(2), 120-129.

Fleisher, A. S., Houston, W. S., Eyler, L. T., Frye, S., Jenkins, C., Thal, L. J., & Bondi, M. W. (2005). Identification of Alzheimer disease risk by functional magnetic resonance imaging. *Arch Neurol*, 62(12), 1881-1888.

Flicker, C., Ferris, S. H., & Reisberg, B. (1991). Mild cognitive impairment in the elderly: predictors of dementia. *Neurology*, 41(7), 1006-1009.

Flood, D. G., & Coleman, P. D. (1988). Neuron numbers and sizes in aging brain: comparisons of human, monkey, and rodent data. *Neurobiol Aging*, 9(5-6), 453-463.

Foster, T. C. (2007). Calcium homeostasis and modulation of synaptic plasticity in the aged brain. *Aging Cell*, 6(3), 319-325.

Freeman, S. H., Raju, S., Hyman, B. T., Frosch, M. P., & Irizarry, M. C. (2007). Plasma Abeta levels do not reflect brain Abeta levels. *J Neuropathol Exp Neurol*, 66(4), 264-271.

Friocourt, G., Poirier, K., Rakic, S., Parnavelas, J. G., & Chelly, J. (2006). The role of ARX in cortical development. *Eur J Neurosci*, 23(4), 869-876.

Galvin, J. E., Roe, C. M., Xiong, C., & Morris, J. C. (2006). Validity and reliability of the AD8 informant interview in dementia. *Neurology*, 67(11), 1942-1948.

Gauthier, S., Reisberg, B., Zaudig, M., Petersen, R. C., Ritchie, K., Broich, K., Belleville, S., Brodaty, H., Bennett, D., Chertkow, H., Cummings, J. L., de Leon, M., Feldman, H., Ganguli, M., Hampel, H., Scheltens, P., Tierney, M. C., Whitehouse, P., & Winblad, B. (2006). Mild cognitive impairment. *Lancet*, 367(9518), 1262-1270.

Geddes, J. W., Tekirian, T. L., Soultanian, N. S., Ashford, J. W., Davis, D. G., & Markesbery, W. R. (1997). Comparison of neuropathologic criteria for the diagnosis of Alzheimer's disease. *Neurobiol Aging*, 18(4 Suppl), S99-105.

Gervais, F., Paquette, J., Morissette, C., Krzywkowski, P., Yu, M., Azzi, M., Lacombe, D., Kong, X., Aman, A., Laurin, J., Szarek, W. A., & Tremblay, P. (2007). Targeting soluble Abeta peptide with Tramiprosate for the treatment of brain amyloidosis. *Neurobiol Aging*, 28(4), 537-547.

Giedd, J. N., Blumenthal, J., Jeffries, N. O., Rajapakse, J. C., Vaituzis, A. C., Liu, H., Berry, Y. C., Tobin, M., Nelson, J., & Castellanos, F. X. (1999). Development of the human corpus callosum during childhood and adolescence: a longitudinal MRI study. *Prog Neuropsychopharmacol Biol Psychiatry*, 23(4), 571-588.

Giedd, J.N., Blumenthal, J., Jeffries, N. O., Castellanos, F. X., Liu, H., Zijdenbos, A., Paus, T., Evans, A. C., Rapoport, J. L. (1999). Brain development during childhood and adolescence: a longitudinal MRI study. *Nat Neurosciencey*, 2(10), 861-863.

Gogtay, N., Giedd, J. N., Lusk, L., Hayashi, K. M., Greenstein, D., Vaituzis, A. C., Nugent, T. F., 3rd, Herman, D. H., Clasen, L. S., Toga, A. W., Rapoport, J. L., & Thompson, P. M. (2004). Dynamic mapping of human cortical development during childhood through early adulthood. *Proc Natl Acad Sci U S A*, 101(21), 8174-8179.

Golde, T. E. (2006). Disease modifying therapy for AD? *J Neurochem*, 99(3), 689-707.

Goldman, W. P., Price, J. L., Storandt, M., Grant, E. A., McKeel, D. W., Jr., Rubin, E. H., & Morris, J. C. (2001). Absence of cognitive impairment or decline in preclinical Alzheimer's disease. *Neurology*, 56(3), 361-367.

Gomes, R. A., Hampton, C., El-Sabeawy, F., Sabo, S. L., & McAllister, A. K. (2006). The dynamic distribution of TrkB receptors before, during, and after synapse formation between cortical neurons. *J Neurosci*, 26(44), 11487-11500.

Gongidi, V., Ring, C., Moody, M., Brekken, R., Sage, E. H., Rakic, P., & Anton, E. S. (2004). SPARC-like 1 regulates the terminal phase of radial glia-guided migration in the cerebral cortex. *Neuron*, 41(1), 57-69.

Good, C. D., Johnsrude, I. S., Ashburner, J., Henson, R. N., Friston, K. J., & Frackowiak, R. S. (2001). A voxel-based morphometric study of ageing in 465 normal adult human brains. *Neuroimage*, 14(1 Pt 1), 21-36.

Graff-Radford, N. R., Crook, J. E., Lucas, J., Boeve, B. F., Knopman, D. S., Ivnik, R. J., Smith, G. E., Younkin, L. H., Petersen, R. C., & Younkin, S. G. (2007). Association of low plasma Abeta42/Abeta40 ratios with increased imminent risk for mild cognitive impairment and Alzheimer disease. *Arch Neurol*, 64(3), 354-362.

Gralle, M., & Ferreira, S. T. (2007). Structure and functions of the human amyloid precursor protein: the whole is more than the sum of its parts. *Prog Neurobiol*, 82(1), 11-32.

Greenwood, P. M. (2000). The frontal aging hypothesis evaluated. *J Int Neuropsychol Soc*, 6(6), 705-726.

Grober, E., Lipton, R. B., Hall, C., & Crystal, H. (2000). Memory impairment on free and cued selective reminding predicts dementia. *Neurology*, 54(4), 827-832.

Guenette, S., Chang, Y., Hiesberger, T., Richardson, J. A., Eckman, C. B., Eckman, E. A., Hammer, R. E., & Herz, J. (2006). Essential roles for the FE65 amyloid precursor protein-interacting proteins in brain development. *Embo J*, 25(2), 420-431.

Head, D., Snyder, A. Z., Girton, L. E., Morris, J. C., & Buckner, R. L. (2005). Frontal-hippocampal double dissociation between normal aging and Alzheimer's disease. *Cereb Cortex*, 15(6), 732-739.

Head, E., Lott, I. T., Patterson, D., Doran, E., & Haier, R. J. (2007). Possible compensatory events in adult Down syndrome brain prior to the development of Alzheimer disease neuropathology: targets for nonpharmacological intervention. *J Alzheimers Dis*, 11(1), 61-76.

Ikonomovic, M. D., Uryu, K., Abrahamson, E. E., Ciallella, J. R., Trojanowski, J. Q., Lee, V. M., Clark, R. S., Marion, D. W., Wisniewski, S. R., & DeKosky, S. T. (2004). Alzheimer's pathology in human temporal cortex surgically excised after severe brain injury. *Exp Neurol*, 190(1), 192-203.

Israely, I., Costa, R. M., Xie, C. W., Silva, A. J., Kosik, K. S., & Liu, X. (2004). Deletion of the neuron-specific protein delta-catenin leads to severe cognitive and synaptic dysfunction. *Curr Biol*, 14(18), 1657-1663.

Jack, C. R., Jr., Petersen, R. C., Xu, Y., O'Brien, P. C., Smith, G. E., Ivnik, R. J., Boeve, B. F., Tangalos, E. G., & Kokmen, E. (2000). Rates of hippocampal atrophy correlate with change in clinical status in aging and AD. *Neurology*, 55(4), 484-489.

Jack, C. R., Jr., Petersen, R. C., Xu, Y. C., O'Brien, P. C., Smith, G. E., Ivnik, R. J., Boeve, B. F., Waring, S. C., Tangalos, E. G., & Kokmen, E. (1999). Prediction of AD with MRI-based hippocampal volume in mild cognitive impairment. *Neurology*, 52(7), 1397-1403.

Jack, C. R., Jr., Shiung, M. M., Gunter, J. L., O'Brien, P. C., Weigand, S. D., Knopman, D. S., Boeve, B. F., Ivnik, R. J., Smith, G. E., Cha, R. H., Tangalos, E. G., & Petersen, R. C. (2004). Comparison of different MRI brain atrophy rate measures with clinical disease progression in AD. *Neurology*, 62(4), 591-600.

Jack, C. R., Jr., Shiung, M. M., Weigand, S. D., O'Brien, P. C., Gunter, J. L., Boeve, B. F., Knopman, D. S., Smith, G. E., Ivnik, R. J., Tangalos, E. G., & Petersen, R. C. (2005). Brain atrophy rates predict subsequent clinical conversion in normal elderly and amnestic MCI. *Neurology*, 65(8), 1227-1231.

Jagust, W., Gitcho, A., Sun, F., Kuczynski, B., Mungas, D., & Haan, M. (2006). Brain imaging evidence of preclinical Alzheimer's disease in normal aging. *Ann Neurol*, 59(4), 673-681.

Jellinger, K. A. (1998). The neuropathological diagnosis of Alzheimer disease. *J Neural Transm Suppl*, 53, 97-118.

Jellinger, K. A., & Bancher, C. (1998). Neuropathology of Alzheimer's disease: a critical update. *J Neural Transm Suppl*, 54, 77-95.

Jernigan T.L., & Tallal, P. (1990). Late childhood changes in brain morphology observable with MRI. *Dev Med Child Neurol*, 32(5), 379-385.

Jernigan T,L., Trauner, D. A., Hesselink, J. R., & Tallal, P. A. (1991) Maturation of human cerebrum observed in vivo during adolescence. *Brain* 114, 2037-2049.

Jernigan, T. L., & Gamst, A. C. (2005). Changes in volume with age--consistency and interpretation of observed effects. *Neurobiol Aging*, 26(9), 1271-1274; discussion 1275-1278.

Johnson, K. A., Lopera, F., Jones, K., Becker, A., Sperling, R., Hilson, J., Londono, J., Siegert, I., Arcos, M., Moreno, S., Madrigal, L., Ossa, J., Pineda, N., Ardila, A., Roselli, M., Albert, M. S., Kosik, K. S., & Rios, A. (2001).

Presenilin-1-associated abnormalities in regional cerebral perfusion. *Neurology*, 56(11), 1545-1551.

Karas, G. B., Scheltens, P., Rombouts, S. A., Visser, P. J., van Schijndel, R. A., Fox, N. C., & Barkhof, F. (2004). Global and local gray matter loss in mild cognitive impairment and Alzheimer's disease. *Neuroimage*, 23(2), 708-716.

Karunanayaka, P. R., Holland, S. K., Schmithorst, V. J., Solodkin, A., Chen, E. E., Szaflarski, J. P., & Plante, E. (2007). Age-related connectivity changes in fMRI data from children listening to stories. *Neuroimage*, 34(1), 349-360.

Kato, M., Das, S., Petras, K., Kitamura, K., Morohashi, K., Abuelo, D. N., Barr, M., Bonneau, D., Brady, A. F., Carpenter, N. J., Cipero, K. L., Frisone, F., Fukuda, T., Guerrini, R., Iida, E., Itoh, M., Lewanda, A. F., Nanba, Y., Oka, A., Proud, V. K., Saugier-Veber, P., Schelley, S. L., Selicorni, A., Shaner, R., Silengo, M., Stewart, F., Sugiyama, N., Toyama, J., Toutain, A., Vargas, A. L., Yanazawa, M., Zackai, E. H., & Dobyns, W. B. (2004). Mutations of ARX are associated with striking pleiotropy and consistent genotype-phenotype correlation. *Hum Mutat*, 23(2), 147-159.

Katzman, R. (1993). Education and the prevalence of dementia and Alzheimer's disease. *Neurology*, 43, 13-20.

Kaye, J. A., Swihart, T., Howieson, D., Dame, A., Moore, M. M., Karnos, T., Camicioli, R., Ball, M., Oken, B., & Sexton, G. (1997). Volume loss of the hippocampus and temporal lobe in healthy elderly persons destined to develop dementia. *Neurology*, 48(5), 1297-1304.

Kelley, B. J., & Petersen, R. C. (2007). Alzheimer's disease and mild cognitive impairment. *Neurol Clin*, 25(3), 577-609.

Kidron, D., Black, S. E., Stanchev, P., Buck, B., Szalai, J. P., Parker, J., Szekely, C., & Bronskill, M. J. (1997). Quantitative MR volumetry in Alzheimer's disease. Topographic markers and the effects of sex and education. *Neurology*, 49, 1504-1512.

Komuro, H., & Rakic, P. (1996). Intracellular Ca2+ fluctuations modulate the rate of neuronal migration. *Neuron*, 17(2), 275-285.

Komuro, H., & Rakic, P. (1998). Orchestration of neuronal migration by activity of ion channels, neurotransmitter receptors, and intracellular Ca2+ fluctuations. *J Neurobiol*, 37(1), 110-130.

Kovari, E., Gold, G., Herrmann, F. R., Canuto, A., Hof, P. R., Bouras, C., & Giannakopoulos, P. (2007). Cortical microinfarcts and demyelination affect cognition in cases at high risk for dementia. *Neurology*, 68(12), 927-931.

Kramer, A. F., & Erickson, K. I. (2007). Capitalizing on cortical plasticity: influence of physical activity on cognition and brain function. *Trends Cogn Sci*, 11(8), 342-348.

Kukar, T., Prescott, S., Eriksen, J. L., Holloway, V., Murphy, M. P., Koo, E. H., Golde, T. E., & Nicolle, M. M. (2007). Chronic administration of R-flurbiprofen attenuates learning impairments in transgenic amyloid precursor protein mice. *BMC Neurosci*, 8, 54.

Lemaitre, H., Crivello, F., Grassiot, B., Alperovitch, A., Tzourio, C., & Mazoyer, B. (2005). Age- and sex-related effects on the neuroanatomy of healthy elderly. *Neuroimage*, 26(3), 900-911.

Lessoway, V.A., Schulzer, M.,Wittmann, B. K., Gagnon, F. A., Wilson, R. D. (2008) Ultrasound fetal biometry charts for a North American Caucasian population. *J Clin Ultrasound* 26(9), 433-453.

Letenneur, L., Gilleron, V., Commenges, D., Helmer, C., Orgogozo, J. M., & Dartigues, J. F. (1999). Are sex and educational level independent predictors of dementia and Alzheimer's disease? Incidence data from the PAQUID project. *J Neurol Neurosurg Psychiatry*, 66(2), 177-183.

Liao, Y. C., Liu, R. S., Teng, E. L., Lee, Y. C., Wang, P. N., Lin, K. N., Chung, C. P., & Liu, H. C. (2005). Cognitive reserve: a SPECT study of 132 Alzheimer's disease patients with an education range of 0-19 years. *Dement Geriatr Cogn Disord*, 20(1), 8-14.

Lim, K. O., Zipursky, R. B., Watts, M. C., & Pfefferbaum, A. (1992). Decreased gray matter in normal aging: an in vivo magnetic resonance study. *J Gerontol*, 47(1), B26-30.

Lippman, J., & Dunaevsky, A. (2005). Dendritic spine morphogenesis and plasticity. *J Neurobiol*, 64(1), 47-57.

Liu, R. S., Lemieux, L., Bell, G. S., Sisodiya, S. M., Shorvon, S. D., Sander, J. W., & Duncan, J. S. (2003). A longitudinal study of brain morphometrics using quantitative magnetic resonance imaging and difference image analysis. *Neuroimage*, 20(1), 22-33.

Lockhart, A., Lamb, J. R., Osredkar, T., Sue, L. I., Joyce, J. N., Ye, L., Libri, V., Leppert, D., & Beach, T. G. (2007). PIB is a non-specific imaging marker of amyloid-beta (A{beta}) peptide-related cerebral amyloidosis. *Brain*.

Lonze, B. E., & Ginty, D. D. (2002). Function and regulation of CREB family transcription factors in the nervous system. *Neuron*, 35(4), 605-623.

Lorent, K., Overbergh, L., Moechars, D., De Strooper, B., Van Leuven, F., & Van den Berghe, H. (1995). Expression in mouse embryos and in adult mouse brain of three members of the amyloid precursor protein family, of the alpha-2-macroglobulin receptor/low density lipoprotein receptor-related protein and of its ligands apolipoprotein E, lipoprotein lipase, alpha-2-macroglobulin and the 40,000 molecular weight receptor-associated protein. *Neuroscience*, 65(4), 1009-1025.

Lovell, M. A., Xie, C., & Markesbery, W. R. (1998). Decreased glutathione transferase activity in brain and ventricular fluid in Alzheimer's disease. *Neurology*, 51(6), 1562-1566.

Mahncke, H. W., Connor, B. B., Appelman, J., Ahsanuddin, O. N., Hardy, J. L., Wood, R. A., Joyce, N. M., Boniske, T., Atkins, S. M., & Merzenich, M. M. (2006). Memory enhancement in healthy older adults using a brain plasticity-based training program: a randomized, controlled study. *Proc Natl Acad Sci U S A*, 103(33), 12523-12528.

Malaterre, J., Ramsay, R. G., & Mantamadiotis, T. (2007). Wnt-Frizzled signalling and the many paths to neural development and adult brain homeostasis. *Front Biosci*, 12, 492-506.

Mally, J., & Stone, T. W. (2007). New advances in the rehabilitation of CNS diseases applying rTMS. *Expert Rev Neurother*, 7(2), 165-177.

Marjaux, E., Hartmann, D., & De Strooper, B. (2004). Presenilins in memory, Alzheimer's disease, and therapy. *Neuron*, 42(2), 189-192.

Mark, V. W., Taub, E., & Morris, D. M. (2006). Neuroplasticity and constraint-induced movement therapy. *Eura Medicophys*, 42(3), 269-284.

Markesbery, W. R., Schmitt, F. A., Kryscio, R. J., Davis, D. G., Smith, C. D., & Wekstein, D. R. (2006). Neuropathologic substrate of mild cognitive impairment. *Arch Neurol*, 63(1), 38-46.

Markus, T. M., Tsai, S. Y., Bollnow, M. R., Farrer, R. G., O'Brien, T. E., Kindler-Baumann, D. R., Rausch, M., Rudin, M., Wiessner, C., Mir, A. K., Schwab, M. E., & Kartje, G. L. (2005). Recovery and brain reorganization after stroke in adult and aged rats. *Ann Neurol*, 58(6), 950-953.

Marui, W., Iseki, E., Kato, M., Akatsu, H., & Kosaka, K. (2004). Pathological entity of dementia with Lewy bodies and its differentiation from Alzheimer's disease. *Acta Neuropathol (Berl)*, 108(2), 121-128.

Masliah, E., Terry, R. D., DeTeresa, R. M., & Hansen, L. A. (1989). Immunohistochemical quantification of the synapse-related protein synaptophysin in Alzheimer disease. *Neurosci Lett*, 103(2), 234-239.

Masur, D. M., Sliwinski, M., Lipton, R. B., Blau, A. D., & Crystal, H. A. (1994). Neuropsychological prediction of dementia and the absence of dementia in healthy elderly persons. *Neurology*, 44(8), 1427-1432.

Mattson, M. P. (2007). Calcium and neurodegeneration. *Aging Cell*, 6(3), 337-350.

McDowell, I., Xi, G., Lindsay, J., & Tierney, M. (2007). Mapping the connections between education and dementia. *J Clin Exp Neuropsychol*, 29(2), 127-141.

McKhann, G., Drachman, D., Folstein, M., Katzman, R., Price, D., & Stadlan, E. M. (1984). Clinical diagnosis of Alzheimer's disease: report of the NINCDS-

ADRDA Work Group under the auspices of Department of Health and Human Services Task Force on Alzheimer's Disease. *Neurology*, 34(7), 939-944.

Metzler-Baddeley, C. (2007). A review of cognitive impairments in dementia with Lewy bodies relative to Alzheimer's disease and Parkinson's disease with dementia. *Cortex*, 43(5), 583-600.

Mileusnic, R., Lancashire, C. L., & Rose, S. P. (2005). Amyloid precursor protein: from synaptic plasticity to Alzheimer's disease. *Ann N Y Acad Sci*, 1048, 149-165.

Miyasaka, T., Watanabe, A., Saito, Y., Murayama, S., Mann, D. M., Yamazaki, M., Ravid, R., Morishima-Kawashima, M., Nagashima, K., & Ihara, Y. (2005). Visualization of newly deposited tau in neurofibrillary tangles and neuropil threads. *J Neuropathol Exp Neurol*, 64(8), 665-674.

Morris, J. C. (1997). Clinical dementia rating: a reliable and valid diagnostic and staging measure for dementia of the Alzheimer type. *Int Psychogeriatr*, 9 Suppl 1, 173-176; discussion 177-178.

Morris, J. C., McKeel, D. W., Jr., Storandt, M., Rubin, E. H., Price, J. L., Grant, E. A., Ball, M. J., & Berg, L. (1991). Very mild Alzheimer's disease: informant-based clinical, psychometric, and pathologic distinction from normal aging. *Neurology*, 41(4), 469-478.

Morris, J. C., & Price, A. L. (2001). Pathologic correlates of nondemented aging, mild cognitive impairment, and early-stage Alzheimer's disease. *J Mol Neurosci*, 17(2), 101-118.

Mortimer, J. A. (1997). Brain reserve and the clinical expression of Alzheimer's disease. *Geriatrics*, 52 Suppl 2, S50-53.

Mortimer, J. A., Borenstein, A. R., Gosche, K. M., & Snowdon, D. A. (2005). Very early detection of Alzheimer neuropathology and the role of brain reserve in modifying its clinical expression. *J Geriatr Psychiatry Neurol*, 18(4), 218-223.

Mosconi, L., Herholz, K., Prohovnik, I., Nacmias, B., De Cristofaro, M. T., Fayyaz, M., Bracco, L., Sorbi, S., & Pupi, A. (2005). Metabolic interaction between ApoE genotype and onset age in Alzheimer's disease: implications for brain reserve. *J Neurol Neurosurg Psychiatry*, 76(1), 15-23.

Mothet, J. P., Rouaud, E., Sinet, P. M., Potier, B., Jouvenceau, A., Dutar, P., Videau, C., Epelbaum, J., & Billard, J. M. (2006). A critical role for the glial-derived neuromodulator D-serine in the age-related deficits of cellular mechanisms of learning and memory. *Aging Cell*, 5(3), 267-274.

Mueller, E. A., Moore, M. M., Kerr, D. C., Sexton, G., Camicioli, R. M., Howieson, D. B., Quinn, J. F., & Kaye, J. A. (1998). Brain volume preserved in healthy elderly through the eleventh decade. *Neurology*, 51(6), 1555-1562.

Nagata, K., Basugi, N., Fukushima, T., Tango, T., Suzuki, I., Kaminuma, T., & Kurashina, S. (1987). A quantitative study of physiological cerebral atrophy with aging. A statistical analysis of the normal range. *Neuroradiology*, 29(4), 327-332.

Newell, K. L., Hyman, B. T., Growdon, J. H., & Hedley-Whyte, E. T. (1999). Application of the National Institute on Aging (NIA)-Reagan Institute criteria for the neuropathological diagnosis of Alzheimer disease. *J Neuropathol Exp Neurol*, 58(11), 1147-1155.

Noctor, S,C., Martinez-Cerdeno, V., & Kriegstein, A. R. (2007). Contribution of intermediate progenitor cells to cortical histogenesis. *Arch Neurol* 64(5), 639-642.

Nordberg, A. (2007). Amyloid imaging in Alzheimer's disease. *Curr Opin Neurol*, 20(4), 398-402.

Ohm, T. G., Muller, H., Braak, H., & Bohl, J. (1995). Close-meshed prevalence rates of different stages as a tool to uncover the rate of Alzheimer's disease-related neurofibrillary changes. *Neuroscience*, 64(1), 209-217.

Olsson, A., Csajbok, L., Ost, M., Hoglund, K., Nylen, K., Rosengren, L., Nellgard, B., & Blennow, K. (2004). Marked increase of beta-amyloid(1-42) and amyloid precursor protein in ventricular cerebrospinal fluid after severe traumatic brain injury. *J Neurol*, 251(7), 870-876.

Papadopoulos, C. M., Tsai, S. Y., Alsbiei, T., O'Brien, T. E., Schwab, M. E., & Kartje, G. L. (2002). Functional recovery and neuroanatomical plasticity following middle cerebral artery occlusion and IN-1 antibody treatment in the adult rat. *Ann Neurol*, 51(4), 433-441.

Pennanen, C., Testa, C., Laakso, M. P., Hallikainen, M., Helkala, E. L., Hanninen, T., Kivipelto, M., Kononen, M., Nissinen, A., Tervo, S., Vanhanen, M., Vanninen, R., Frisoni, G. B., & Soininen, H. (2005). A voxel based morphometry study on mild cognitive impairment. *J Neurol Neurosurg Psychiatry*, 76(1), 11-14.

Perneczky, R., Diehl-Schmid, J., Drzezga, A., & Kurz, A. (2007). Brain reserve capacity in frontotemporal dementia: a voxel-based (18)F-FDG PET study. *Eur J Nucl Med Mol Imaging*, 34(7), 1082-1087.

Perneczky, R., Diehl-Schmid, J., Pohl, C., Drzezga, A., & Kurz, A. (2007). Non-fluent progressive aphasia: cerebral metabolic patterns and brain reserve. *Brain Res*, 1133(1), 178-185.

Perneczky, R., Drzezga, A., Diehl-Schmid, J., Schmid, G., Wohlschlager, A., Kars, S., Grimmer, T., Wagenpfeil, S., Monsch, A., & Kurz, A. (2006). Schooling mediates *Brain Res*erve in Alzheimer's disease: findings of fluoro-deoxy-glucose-positron emission tomography. *J Neurol Neurosurg Psychiatry*, 77(9), 1060-1063.

Petersen, R. C. (1998). Clinical subtypes of Alzheimer's disease. *Dement Geriatr Cogn Disord*, 9 Suppl 3, 16-24.

Petersen, R. C., Parisi, J. E., Dickson, D. W., Johnson, K. A., Knopman, D. S., Boeve, B. F., Jicha, G. A., Ivnik, R. J., Smith, G. E., Tangalos, E. G., Braak, H., & Kokmen, E. (2006). Neuropathologic features of amnestic mild cognitive impairment. *Arch Neurol*, 63(5), 665-672.

Petersen, R. C., Smith, G. E., Waring, S. C., Ivnik, R. J., Tangalos, E. G., & Kokmen, E. (1999). Mild cognitive impairment: clinical characterization and outcome. *Arch Neurol*, 56(3), 303-308.

Pfefferbaum, A., Mathalon, D. H., Sullivan, E. V., Rawles, J. M., Zipursky, R. B., & Lim, K. O. (1994). A quantitative magnetic resonance imaging study of changes in brain morphology from infancy to late adulthood. *Arch Neurol*, 51(9), 874-887.

Pfefferbaum, A., Sullivan, E. V., Swan, G. E., & Carmelli, D. (2000). Brain structure in men remains highly heritable in the seventh and eighth decades of life. *Neurobiol Aging*, 21(1), 63-74.

Plassman, B. L., Welsh, K. A., Helms, M., Brandt, J., Page, W. F., & Breitner, J. C. S. (1995). Intelligence and education as predictors of cognitive state in late life: a 50-year follow-up. *Neurology*, 45, 1446-1450.

Potter, G. G., Plassman, B. L., Helms, M. J., Foster, S. M., & Edwards, N. W. (2006). Occupational characteristics and cognitive performance among elderly male twins. *Neurology*, 67(8), 1377-1382.

Price, D. L., Thinakaran, G., Borchelt, D. R., Martin, L. J., Crain, B. J., Sisodia, S. S., & Troncoso, J. C. (1998). Neuropathology of Alzheimer's disease and animal models. In W. R. Markesbery (Ed.), *Neuropathology of dementing disorders*. London: Arnold.

Rakic, P. (1990). Principles of neural cell migration. *Experientia*, 46(9), 882-891.

Rakic, P. (2007). The radial edifice of cortical architecture: From neuronal silhouettes to genetic engineering. *Brain Res Rev*.

Ramachandran, V. S. (2005a). Plasticity and functional recovery in neurology. *Clin Med*, 5(4), 368-373.

Ramachandran, V. S. (2005b). *Plasticity and functional recovery in neurology*. (1470-2118 (Print)).

Ramachandran, V. S., & Rogers-Ramachandran, D. (2000). Phantom limbs and neural plasticity. *Arch Neurol*, 57(3), 317-320.

Ramic, M., Emerick, A. J., Bollnow, M. R., O'Brien, T. E., Tsai, S. Y., & Kartje, G. L. (2006). Axonal plasticity is associated with motor recovery following amphetamine treatment combined with rehabilitation after brain injury in the adult rat. *Brain Res*, 1111(1), 176-186.

Rauch, R. A., & Jinkins, J. R. (1994). Analysis of cross-sectional area measurements of the corpus callosum adjusted for brain size in male and female subjects from childhood to adulthood. *Behav Brain Res*, 64(1-2), 65-78.

Raz, N., Gunning, F. M., Head, D., Dupuis, J. H., McQuain, J., Briggs, S. D., Loken, W. J., Thornton, A. E., & Acker, J. D. (1997). Selective aging of the human cerebral cortex observed in vivo: differential vulnerability of the prefrontal gray matter. *Cereb Cortex*, 7(3), 268-282.

Reiman, E. M., Caselli, R. J., Yun, L. S., Kewei, C., Bandy, D., Minishima, S., Thibideau, S. N., & Osborne, D. (1996). Preclinical evidence of Alzheimer's disease in persons homozygous for the e4 allele for apolipoprotein E. *NEJM*, 334, 752-758.

Reiman, E. M., Webster, J. A., Myers, A. J., Hardy, J., Dunckley, T., Zismann, V. L., Joshipura, K. D., Pearson, J. V., Hu-Lince, D., Huentelman, M. J., Craig, D. W., Coon, K. D., Liang, W. S., Herbert, R. H., Beach, T., Rohrer, K. C., Zhao, A. S., Leung, D., Bryden, L., Marlowe, L., Kaleem, M., Mastroeni, D., Grover, A., Heward, C. B., Ravid, R., Rogers, J., Hutton, M. L., Melquist, S., Petersen, R. C., Alexander, G. E., Caselli, R. J., Kukull, W., Papassotiropoulos, A., & Stephan, D. A. (2007). GAB2 alleles modify Alzheimer's risk in APOE epsilon4 carriers. *Neuron*, 54(5), 713-720.

Resnick, S. M., Pham, D. L., Kraut, M. A., Zonderman, A. B., & Davatzikos, C. (2003). Longitudinal magnetic resonance imaging studies of older adults: a shrinking brain. *J Neurosci*, 23(8), 3295-3301.

Roe, C. M., Xiong, C., Miller, J. P., & Morris, J. C. (2007). Education and Alzheimer disease without dementia: support for the cognitive reserve hypothesis. *Neurology*, 68(3), 223-228.

Rovio, S., Kareholt, I., Viitanen, M., Winblad, B., Tuomilehto, J., Soininen, H., Nissinen, A., & Kivipelto, M. (2007). Work-related physical activity and the risk of dementia and Alzheimer's disease. *Int J Geriatr Psychiatry*, 22(9), 874-882.

Rowe, C. C., Ng, S., Ackermann, U., Gong, S. J., Pike, K., Savage, G., Cowie, T. F., Dickinson, K. L., Maruff, P., Darby, D., Smith, C., Woodward, M., Merory, J., Tochon-Danguy, H., O'Keefe, G., Klunk, W. E., Mathis, C. A.,

Price, J. C., Masters, C. L., & Villemagne, V. L. (2007). Imaging beta-amyloid burden in aging and dementia. *Neurology*, 68(20), 1718-1725.

Salat, D. H., Kaye, J. A., & Janowsky, J. S. (1999). Prefrontal gray and white matter volumes in healthy aging and Alzheimer disease. *Arch Neurol*, 56(3), 338-344.

Sarkisian, M. R., Bartley, C. M., Chi, H., Nakamura, F., Hashimoto-Torii, K., Torii, M., Flavell, R. A., & Rakic, P. (2006). MEKK4 signaling regulates filamin expression and neuronal migration. *Neuron*, 52(5), 789-801.

Saura, C. A., Choi, S. Y., Beglopoulos, V., Malkani, S., Zhang, D., Shankaranarayana Rao, B. S., Chattarji, S., Kelleher, R. J., 3rd, Kandel, E. R., Duff, K., Kirkwood, A., & Shen, J. (2004). Loss of presenilin function causes impairments of memory and synaptic plasticity followed by age-dependent neurodegeneration. *Neuron*, 42(1), 23-36.

Scahill, R. I., Frost, C., Jenkins, R., Whitwell, J. L., Rossor, M. N., & Fox, N. C. (2003). A longitudinal study of brain volume changes in normal aging using serial registered magnetic resonance imaging. *Arch Neurol*, 60(7), 989-994.

Scarmeas, N., Zarahn, E., Anderson, K. E., Honig, L. S., Park, A., Hilton, J., Flynn, J., Sackeim, H. A., & Stern, Y. (2004). Cognitive reserve-mediated modulation of positron emission tomographic activations during memory tasks in Alzheimer disease. *Arch Neurol*, 61(1), 73-78.

Scheff, S. W., & Price, D. A. (2006). Alzheimer's disease-related alterations in synaptic density: neocortex and hippocampus. *J Alzheimers Dis*, 9(3 Suppl), 101-115.

Scheff, S. W., Price, D. A., Schmitt, F. A., & Mufson, E. J. (2006). Hippocampal synaptic loss in early Alzheimer's disease and mild cognitive impairment. *Neurobiol Aging*, 27(10), 1372-1384.

Scherder, E., Eggermont, L., Sergeant, J., & Boersma, F. (2007). Physical activity and cognition in Alzheimer's disease: relationship to vascular risk factors, executive functions and gait. *Rev Neurosci*, 18(2), 149-158.

Schmidt, C., Lepsverdize, E., Chi, S. L., Das, A. M., Pizzo, S. V., Dityatev, A., & Schachner, M. (2007). Amyloid precursor protein and amyloid beta-peptide bind to ATP synthase and regulate its activity at the surface of neural cells. *Mol Psychiatry*.

Schmithorst, V. J., Holland, S. K., & Dardzinski, B. J. (2007). Developmental differences in white matter architecture between boys and girls. *Hum Brain Mapp*.

Schmithorst, V. J., Holland, S. K., & Plante, E. (2006). Object identification and lexical/semantic access in children: A functional magnetic resonance imaging study of word-picture matching. *Hum Brain Mapp*.

Schmithorst, V. J., Wilke, M., Dardzinski, B. J., & Holland, S. K. (2002). Correlation of white matter diffusivity and anisotropy with age during childhood and adolescence: a cross-sectional diffusion-tensor MR imaging study. *Radiology*, 222(1), 212-218.

Schmitt, F. A., Davis, D. G., Wekstein, D. R., Smith, C. D., Ashford, J. W., & Markesbery, W. R. (2000). "Preclinical" AD revisited: neuropathology of cognitively normal older adults. *Neurology*, 55(3), 370-376.

Schneider, J. A., Arvanitakis, Z., Bang, W., & Bennett, D. A. (2007). Mixed brain pathologies account for most dementia cases in community-dwelling older persons. *Neurology*.

Schofield, P. W., Mosesson, R. E., Stern, Y., & Mayeux, R. (1995). The age at onset of Alzheimer's disease and an intracranial area measurement. A relationship. *Arch.Neurol.*, 52, 95-98.

Sharma, K., Fong, D. K., & Craig, A. M. (2006). Postsynaptic protein mobility in dendritic spines: long-term regulation by synaptic NMDA receptor activation. *Mol Cell Neurosci*, 31(4), 702-712.

Sidman, R. L., & Rakic, P. (1982). Development of the human central nervous system. In W. Haymaker & R. D. Adams (Eds.), *Histology and Histopathology of the Nervous System* (pp. 1-112). Springfield Ill.: Thomas.

Silva, A. J., & Giese, K. P. (1994). Plastic genes are in! *Curr Opin Neurobiol*, 4(3), 413-420.

Small, D. H. (2004). Mechanisms of synaptic homeostasis in Alzheimer's disease. *Curr Alzheimer Res*, 1(1), 27-32.

Small, G. W., Ercoli, L. M., Silverman, D. H., Huang, S. C., Komo, S., Bookheimer, S. Y., Lavretsky, H., Miller, K., Siddarth, P., Rasgon, N. L., Mazziotta, J. C., Saxena, S., Wu, H. M., Mega, M. S., Cummings, J. L., Saunders, A. M., Pericak-Vance, M. A., Roses, A. D., Barrio, J. R., & Phelps, M. E. (2000). Cerebral metabolic and cognitive decline in persons at genetic risk for Alzheimer's disease. *Proc Natl Acad Sci U S A*, 97(11), 6037-6042.

Small, S. A., Stern, Y., Tang, M., & Mayeux, R. (1999). Selective decline in memory function among healthy elderly. *Neurology*, 52(7), 1392-1396.

Smith, C. D., Andersen, A. H., Kryscio, R. J., Schmitt, F. A., Blonder, L. X., Kindy, M. S., & Avison, M. J. (1999). Altered brain activation in normal subjects at risk for Alzheimer's disease. *Neurology*, 53, 1391-1396.

Smith, C. D., Andersen, A. H., Kryscio, R. J., Schmitt, F. A., Kindy, M. S., Blonder, L. X., & Avison, M. J. (1999). Altered brain activation in cognitively intact individuals at high risk for Alzheimer's disease. *Neurology*, 53(7), 1391-1396.

Smith, C. D., Andersen, A. H., Kryscio, R. J., Schmitt, F. A., Kindy, M. S., Blonder, L. X., & Avison, M. J. (2002). Women at risk for AD show increased parietal activation during a fluency task. *Neurology, 58*(8), 1197-1202.

Smith, C. D., Chebrolu, H., Wekstein, D. R., Schmitt, F. A., Jicha, G. A., Cooper, G., & Markesbery, W. R. (2007). Brain structural alterations before mild cognitive impairment. *Neurology, 68*(16), 1268-1273.

Smith, C. D., Chebrolu, H., Wekstein, D. R., Schmitt, F. A., & Markesbery, W. R. (2007). Age and gender effects on human brain anatomy: a voxel-based morphometric study in healthy elderly. *Neurobiol Aging, 28*(7), 1075-1087.

Smith, C. D., Kryscio, R. J., Schmitt, F. A., Lovell, M. A., Blonder, L. X., Rayens, W. S., & Andersen, A. H. (2005). Longitudinal functional alterations in asymptomatic women at risk for Alzheimer's disease. *J Neuroimaging, 15*(3), 271-277.

Smith, D. E., Roberts, J., Gage, F. H., & Tuszynski, M. H. (1999). Age-associated neuronal atrophy occurs in the primate brain and is reversible by growth factor gene therapy. *Proc Natl Acad Sci U S A, 96*(19), 10893-10898.

Smyth, K. A., Fritsch, T., Cook, T. B., McClendon, M. J., Santillan, C. E., & Friedland, R. P. (2004). Worker functions and traits associated with occupations and the development of AD. *Neurology, 63*(3), 498-503.

Snowdon, D. A., Greiner, L. H., Mortimer, J. A., Riley, K. P., Greiner, P. A., & Markesbery, W. R. (1997). Brain infarction and the clinical expression of Alzheimer disease. The Nun Study. *Jama, 277*(10), 813-817.

Snowdon, D. A., Kemper, S. J., Mortimer, J. A., Greiner, L. H., Wekstein, D. R., & Markesbery, W. R. (1996). Linguistic ability in early life and cognitive function and Alzheimer's disease in late life. Findings from the Nun Study . *JAMA, 275*, 528-532.

Sowell, E.R., Thompson, P. M., Tessner, K. D., & Toga, A. W. (2001). Mapping continued brain growth and gray matter density reduction in dorsal frontal cortex: Inverse relationships during postadolescent brain maturation. *J Neurosci* 21(22), 8819-8829.

Sperling, R. (2007). Functional MRI studies of associative encoding in normal aging, mild cognitive impairment, and Alzheimer's disease. *Ann N Y Acad Sci, 1097*, 146-155.

Stefanova, E., Nilsson, A., Andersson, J., Axelman, K., Langstrom, B., Viitanen, M., Lannfelt, L., & Nordberg, A. (2002). Abnormalities in cerebral glucose metabolism of familial Alzheimer's disease associated with 670/671 APP mutation and presenilin 1 gene mutations. *Paper presented at the ICAD*, Kyoto, Japan.

Steiner, B., Wolf, S., & Kempermann, G. (2006). Adult neurogenesis and neurodegenerative disease. *Regen Med*, 1(1), 15-28.

Stern, R. A., Silva, S. G., Chaisson, N., & Evans, D. L. (1996). Influence of cognitive reserve on neuropsychological functioning in asymptomatic human immunodeficiency virus-1 infection. *Arch Neurol*, 53(2), 148-153.

Stern, Y. (2002). What is cognitive reserve? Theory and research application of the reserve concept. *J Int Neuropsychol Soc*, 8(3), 448-460.

Stern, Y., Alexander, G. E., Prohovnik, I., & Mayeux, R. (1992). Inverse relationship between education and parietotemporal perfusion deficit in Alzheimer's disease. *Ann Neurol*, 32(3), 371-375.

Stern, Y., Gurland, B., Tatemichi, T. K., Tang, M. X., Wilder, D., & Mayeux, R. (1994). Influence of education and occupation on the incidence of ALzheimer's disease. *JAMA*, 13, 1004-1010.

Stettler, D. D., Yamahachi, H., Li, W., Denk, W., & Gilbert, C. D. (2006). Axons and synaptic boutons are highly dynamic in adult visual cortex. *Neuron*, 49(6), 877-887.

Storandt, M., Grant, E. A., Miller, J. P., & Morris, J. C. (2006). Longitudinal course and neuropathologic outcomes in original vs revised MCI and in pre-MCI. *Neurology*, 67(3), 467-473.

Stromme, P., Mangelsdorf, M. E., Shaw, M. A., Lower, K. M., Lewis, S. M., Bruyere, H., Lutcherath, V., Gedeon, A. K., Wallace, R. H., Scheffer, I. E., Turner, G., Partington, M., Frints, S. G., Fryns, J. P., Sutherland, G. R., Mulley, J. C., & Gecz, J. (2002). Mutations in the human ortholog of Aristaless cause X-linked mental retardation and epilepsy. *Nat Genet*, 30(4), 441-445.

Sunderland, A., & Tuke, A. (2005). Neuroplasticity, learning and recovery after stroke: a critical evaluation of constraint-induced therapy. *Neuropsychol Rehabil*, 15(2), 81-96.

Sweatt, J. D. (2001). The neuronal MAP kinase cascade: a biochemical signal integration system subserving synaptic plasticity and memory. *J Neurochem*, 76(1), 1-10.

Sweatt, J. D., Weeber, E. J., & Lombroso, P. J. (2003). Genetics of childhood disorders: LI. Learning and memory, Part 4: Human cognitive disorders and the ras/ERK/CREB pathway. *J Am Acad Child Adolesc Psychiatry*, 42(6), 741-744.

Szaflarski, J. P., Holland, S. K., Schmithorst, V. J., & Byars, A. W. (2006). fMRI study of language lateralization in children and adults. *Hum Brain Mapp*, 27(3), 202-212.

Szczygielski, J., Mautes, A., Steudel, W. I., Falkai, P., Bayer, T. A., & Wirths, O. (2005). Traumatic brain injury: cause or risk of Alzheimer's disease? A review of experimental studies. *J Neural Transm*, 112(11), 1547-1564.

Tao-Cheng, J. H. (2006). Activity-related redistribution of presynaptic proteins at the active zone. *Neuroscience*, 141(3), 1217-1224.

Terry, R. D. (2006). Alzheimer's disease and the aging brain. *J Geriatr Psychiatry Neurol*, 19(3), 125-128.

Thomas, G. M., & Huganir, R. L. (2004). MAPK cascade signalling and synaptic plasticity. *Nat Rev Neurosci*, 5(3), 173-183.

Tierney, M. C., Yao, C., Kiss, A., & McDowell, I. (2005). Neuropsychological tests accurately predict incident Alzheimer disease after 5 and 10 years. *Neurology*, 64(11), 1853-1859.

Trivedi, M. A., Schmitz, T. W., Ries, M. L., Torgerson, B. M., Sager, M. A., Hermann, B. P., Asthana, S., & Johnson, S. C. (2006). Reduced hippocampal activation during episodic encoding in middle-aged individuals at genetic risk of Alzheimer's disease: a cross-sectional study. *BMC Med*, 4, 1.

Tsanov, M., & Manahan-Vaughan, D. (2007). The adult visual cortex expresses dynamic synaptic plasticity that is driven by the light/dark cycle. *J Neurosci*, 27(31), 8414-8421.

Turner, P. R., O'Connor, K., Tate, W. P., & Abraham, W. C. (2003). Roles of amyloid precursor protein and its fragments in regulating neural activity, plasticity and memory. *Prog Neurobiol*, 70(1), 1-32.

Unverzagt, F. W., Hui, S. L., Farlow, M. R., Hall, K. S., & Hendrie, H. C. (1998). Cognitive decline and education in mild dementia. *Neurology*, 50(1), 181-185.

Uylings, H. B., & de Brabander, J. M. (2002). Neuronal changes in normal human aging and Alzheimer's disease. *Brain Cogn*, 49(3), 268-276.

Van der Borght, K., Havekes, R., Bos, T., Eggen, B. J., & Van der Zee, E. A. (2007). Exercise improves memory acquisition and retrieval in the Y-maze task: relationship with hippocampal neurogenesis. *Behav Neurosci*, 121(2), 324-334.

Walhovd, K. B., Fjell, A. M., Reinvang, I., Lundervold, A., Dale, A. M., Eilertsen, D. E., Quinn, B. T., Salat, D., Makris, N., & Fischl, B. (2005). Effects of age on volumes of cortex, white matter and subcortical structures. *Neurobiol Aging*, 26(9), 1261-1270; discussion 1275-1268.

Walker-Batson, D., Curtis, S., Natarajan, R., Ford, J., Dronkers, N., Salmeron, E., Lai, J., & Unwin, D. H. (2001). A double-blind, placebo-controlled study of the use of amphetamine in the treatment of aphasia. *Stroke*, 32(9), 2093-2098.

Walsh, D. M., & Selkoe, D. J. (2004). Deciphering the molecular basis of memory failure in Alzheimer's disease. *Neuron*, 44(1), 181-193.

Waltereit, R., & Weller, M. (2003). Signaling from cAMP/PKA to MAPK and synaptic plasticity. *Mol Neurobiol*, 27(1), 99-106.

Weisman, D., Cho, M., Taylor, C., Adame, A., Thal, L. J., & Hansen, L. A. (2007). In dementia with Lewy bodies, Braak stage determines phenotype, not Lewy body distribution. *Neurology*, 69(4), 356-359.

Wilke, M., Krageloh-Mann, I., & Holland, S. K. (2007). Global and local development of gray and white matter volume in normal children and adolescents. *Exp Brain Res*, 178(3), 296-307.

Wilson, R. S., Bennett, D. A., Bienias, J. L., Aggarwal, N. T., Mendes De Leon, C. F., Morris, M. C., Schneider, J. A., & Evans, D. A. (2002). Cognitive activity and incident AD in a population-based sample of older persons. *Neurology*, 59(12), 1910-1914.

Wines-Samuelson, M., & Shen, J. (2005). Presenilins in the developing, adult, and aging cerebral cortex. *Neuroscientist*, 11(5), 441-451.

Wishart, H. A., Saykin, A. J., McAllister, T. W., Rabin, L. A., McDonald, B. C., Flashman, L. A., Roth, R. M., Mamourian, A. C., Tsongalis, G. J., & Rhodes, C. H. (2006). Regional brain atrophy in cognitively intact adults with a single APOE epsilon4 allele. *Neurology*, 67(7), 1221-1224.

Wishart, H. A., Saykin, A. J., Rabin, L. A., Santulli, R. B., Flashman, L. A., Guerin, S. J., Mamourian, A. C., Belloni, D. R., Rhodes, C. H., & McAllister, T. W. (2006). Increased brain activation during working memory in cognitively intact adults with the APOE epsilon4 allele. *Am J Psychiatry*, 163(9), 1603-1610.

Wisniewski, H. M., & Silverman, W. (1997). Diagnostic criteria for the neuropathological assessment of Alzheimer's disease: current status and major issues. *Neurobiol Aging*, 18(4 Suppl), S43-50.

Wolf, H., Julin, P., Gertz, H. J., Winblad, B., & Wahlund, L. O. (2004). Intracranial volume in mild cognitive impairment, Alzheimer's disease and vascular dementia: evidence for Brain Reserve? *Int J Geriatr Psychiatry*, 19(10), 995-1007.

Xu, Y., Jack, C. R., Jr., O'Brien, P. C., Kokmen, E., Smith, G. E., Ivnik, R. J., Boeve, B. F., Tangalos, R. G., & Petersen, R. C. (2000). Usefulness of MRI measures of entorhinal cortex versus hippocampus in AD. *Neurology*, 54(9), 1760-1767.

Y Cajal, S. R. (1995). *Histology of the nervous system* (S. Swanson & L. W. Swanson, Trans. Vol. 6). New York: Oxford University Press.

Yaffe, K., Petersen, R. C., Lindquist, K., Kramer, J., & Miller, B. (2006). Subtype of mild cognitive impairment and progression to dementia and death. *Dement Geriatr Cogn Disord*, 22(4), 312-319.

Yakovlev, P. I., & Lecours, A. R. (1967). The myelinogenetic cycles of regional maturation of the brain. In A. Mankowski (Ed.), *Regional development of the brain in early life* (pp. 3-69). Philadelphia: Davis.

Zacharia, A., Zimine, S., Lovblad, K. O., Warfield, S., Thoeny, H., Ozdoba, C., Bossi, E., Kreis, R., Boesch, C., Schroth, G., & Huppi, P. S. (2006). Early assessment of brain maturation by MR imaging segmentation in neonates and premature infants. *AJNR Am J Neuroradiol*, 27(5), 972-977.

Zhang, L., Thomas, K. M., Davidson, M. C., Casey, B. J., Heier, L. A., & Ulug, A. M.. (2005) MR quantitation of volume and diffusion changes in the developing brain. *AJNR*. 26(1), 45-49.

Zhang, Y. W., Wang, R., Liu, Q., Zhang, H., Liao, F. F., & Xu, H. (2007). Presenilin/gamma-secretase-dependent processing of beta-amyloid precursor protein regulates EGF receptor expression. *Proc Natl Acad Sci U S A*, 104(25), 10613-10618.

In: Cognitive Sciences Research Progress
Editor: Miao-Kun Sun

ISBN: 978-1-60456-392-4
© 2008 Nova Science Publishers, Inc.

Chapter 4

SEMANTICALLY MEDIATED INTEGRATION OF COGNITION IN *HOMO SAPIENS*: EVOLUTION, GRAMMAR, UNCERTAINTY, AND COGNITIVE ACCURACY

Charles E. Bailey
Medical Director, Accurate Clinical Trials, Inc.

ABSTRACT

This article reviews research on human brain, cognition, language, behavior, and evolution to posit the value of operating with a stable reference point based on cognitive accuracy and a rational bias. Drawing on rational emotive, cognitive behavioral and cognitive neuroscience on the one hand and a general brain model of frontal lobe executive function and working memory on the other, along with proposed language mediation of cognitive processes, this review yields potential implications for maximizing brain functioning of *Homo sapiens*. Cognitive thought processes depend on the operations and interactions of specific brain structures and networks, functioning more effectively under conditions of cognitive accuracy (including accurate information, thought process accuracy, and event-level accuracy). However, typical cognitive processes appear to promote the adoption and use of subjective cultural beliefs, mediated by language and grammatical habits mostly learned during early development. In turn, these grammatical habits tend to bias humans toward cognitive inaccuracies. On the other hand, a process that applies informed frontal lobe executive functioning to the mediation of cognition, emotion, and behavior may help to minimize the negative effects of indiscriminately applied cultural belief systems, provide a naturalistic framework for future research and ultimately enhance cognitive accuracy as a reference point for evaluating humans while offering improved relative environmental homeostasis.

Key words: neuroscience, rational, evolution, grammar, cognition, cultural belief systems.

INTRODUCTION

As an evolving species, *Homo sapiens* tend to use more primitive, inherently inadequate tools to measure the results of thought and behavior. Lacking awareness of how we use words to think and speak, and measuring our success by unquestioned cultural belief systems that we have accepted uncritically, we frequently fall prey to confusion, misunderstanding, and emotional turmoil (Browne & Keeley, 2007, pp. 182-5). Human cognition and behavior developed as a product of evolution, socialization, development, and language mediation, as proposed by Luria, influenced by Vygotsky (Luria, 1981, pp. 1-13) This article contends that application of components of accurate human mental functioning and evaluations, mediated by language, will enhance the probability of more successful, rational outcomes (Bailey, 2006). "Rational" as used here derives from cognitive accuracy and refers to purposeful cognition, evaluation, behavior, learning, and informed deliberation. Rational cognition promotes adaptive, pragmatic, practical, flexible decision-making matched as closely as possible with the present instead of the past. In this view, rationality does not imply finding or knowing a supposed single right answer, but rather recognizes that, lacking omniscience, we do best to prepare for a variety of possible outcomes and adapt readily when things do not happen as we would prefer. "In forming opinion about future events, [rational expectations imply] the use of all available information to assess the probabilities of the possible states of the world. More simply, [rational] expectations [are those] that are as correct as is possible with available information" (Deardorff, 2006). Such flexibility is preferable to "irrationality," used here to refer to behavior that is rigid and reactive, especially cognition oriented towards learned rather than learning behavior, and based on belief rather than evidence, i.e., operating with outdated non-contextual information containing unexamined cognitive inaccuracies that promote incongruent stimulus-bound event-level reactions and primitive objectification.

BRAIN, BONDING, LANGUAGE, CULTURE, AND EVOLUTION

To understand normal human brain functioning and cognition, it may be helpful to distinguish independent variables among organisms, especially between mammals, higher primates, and human primates. What are the variables and how do they affect the species? What does this mean to us as *Homo sapiens*? The largest obvious difference between humans and other primates is the unique prefrontal cortex (PFC) that supports language (Broca, 1877). Subsequent to this distinction, as far as we know, we have more words than any other species (LeDoux, 2002, p. 198). Likewise, words and language make up the largest differences among human populations, forming the basis for relationships and individual cultural belief system values (Luria, 1981, p. 6-7). Humans innately seem to express social attributes for relationships at many levels (Panksepp, 1998, 221-99). Relationships depend on interactions between individuals including cooperation and coordination. These interactions in turn usually depend largely on communication (Cacioppo & Berntson, 2004, p. 978). Faulty, or inaccurate, communication tends to have an adverse effect on relationships at all levels.

Bonding and the Limbic System

We share many of our brain attributes with other primates, particularly our limbic system, which contributes to our lower-level emotional cognition. We can trace the large variation in socialization among species to the differential distribution of oxytocin and arginine-vasopressin (avp) receptors and the pathways that enable bonding, attachment, affiliative behavior, and creation of cohesive groups (Insel, 1997; Bartels & Zeki, 2004; Young & Wang, 2004; Fisher et al, 2002; Cho et al, 1999; Lim et al, 2004). These oxytocin and avp attachment receptors support the bonding cascade that mediates reward pathways in the ventral tegmental area (VTA), demonstrating associations with the nucleus accumbens, bed nucleus of the stria terminalis (BNST), and interstitial nucleus of the posterior limb of the anterior commissure (IPAC) (Heimer et al, 2005, pp. 61-65). These reward pathways add a positive emotional valence, or value, to bonding and affiliative behavior, mediated (at least in part) by a positive shift in VTA dopamine and probably endogenous opioids (Panksepp 1998, p. 263). These pathways appear to overlay lower-level limbic reinforcement and pain pathways identified in mammals and other primates (Tranel, 2000, p. 218). This suggests an evolutionary adaptation of social behavior to preexisting limbic and somatic pain-reward pathways. (See Caldwell & Young, 2006, for a review of oxytocin and vasopressin.)

Unbonding, rejection, social disapproval, or exclusion from social groups (Cacioppo & Berntson, 2004, p. 983) can induce a negative limbic valence, or value. Threats to attachment or approval may produce emotional pain, jealousy, anger, aggression and violence (Insel, 1997). This negative valence appears to correlate with increased amygdala activity. Studies show that the stress induced by detachment or disapproval triggers bereavement-related syndromes, correlated with an increase in stress hormones, corticotrophin releasing factor (CRF), and a decrease in brain-derived neuronal growth factor (BDNF). Other mammals also express increased CRF associated with social defeat and subordination (Ferris, 2006, p. 167-8). This increased CRF in turn correlates with increased amygdala activity and possibly with the downshifting of cognition to implicit limbic and automatic striatal pathways, with decreased PFC and hippocampal volume coupled with diminished executive cognition and attenuated memory. These symptoms echo those of anxiety and depression, sometimes leading to anger and aggression (Panksepp, 1998, p. 205). Shifts in the limbic bonding axis have a dramatic impact on emotional homeostasis, rewarding bonding, affiliation, and approval, while punishing detachment and disapproval. In social animals like human primates, this fundamentally influences cognition, emotion, and behavior (Nair & Young, 2006).

Differentiation, Language, and Socialization

While inter-species social behavior in non-primates and primates may range from very similar to very different, among primates the habit of language distinguishes human primates from our close primate relatives. "Human language offers replication machinery for unlimited cultural evolution", possibly, "representing the biggest invention of the last 600 million years" (Nowak, 2006, p. 249-50). Similar to many social species, humans normally appear to share a genetic predisposition for relationships. For us, however, the innate tendency to bond and affiliate is intimately associated with cultural belief systems, mediated by language (Panksepp, 1998, p. 245). This underscores the importance and impact of language and communication on relationships, thought, emotion, and behavior (Damasio, 2000, p. 17).

Cultural Belief Systems

The human ability for complex language, formulations, thought, and verbal communication helps define our social interactions (Grafman, 2002, p. 298; Mitchell et al., 2006, p. 63; Risberg, 2006, p. 8-7). Words, grammar, and language support cultural belief systems, which form the major independent variable among normal humans (Luria, 1981, pp. 205-9). Normal human brains exhibit relatively consistent form and function across cultures (LeDoux, 2002, p. 231), but cultural belief systems differ, sometimes dramatically (Whorf, 1956, p. 221). We implicitly learn the structure and rules of what we think and how we think from the culture that we grow up in (Nowak, 2006, p. 263). In other words, communication of cultural belief systems depends on hand-me-down semantics and grammar, which vary from culture to culture (Sapir, 1949, p. 162). Language identifies and defines cultural belief systems and represents a distinguishing group characteristic, directly affecting the thoughts, emotions, and behavior of the group (Adolphs, 2006, p. 269-74).

Cultural belief systems and groups frequently overlap or contain subgroups, but a prerequisite set of beliefs usually determines membership. These systems form within social groups due to our inherited propensity for bonding and affiliations. We find unique belief systems at all scales: individuals, small groups such as families or affiliations, and large groups such as entire societies or states. The rules they embody for the group, regarding thought and behavior, usually pass down from elder members and define the particular belief system. Because the rules and beliefs have direct impact on thought and behavior, they also have a dramatic effect on emotions.

Language, Semantics, grammar, and Cultural Variance

We may assume that a person born and raised in a particular cultural belief system would think, feel, and behave differently than one born elsewhere, not because their brains differ but because they internalize different cultural beliefs (Benjafield, 2007, p. 237; Thompson-Schill et al., 2006, pp. 178-9; Vygotsky & Luria, 1993, pp. 230-31). Even though they share basic human emotions, they will react differently to stimuli based on culturally defined values (Browne & Keeley, 2007, p. 53-69). Words culturally bind our thoughts, our beliefs, and subsequently our behaviors and emotions in a wide range of circumstances (Phelps, 2004, p.1008). Since cultural belief systems rely heavily on semantics—the use and meaning of language—one might also conclude that the use of semantics and grammar figure as the largest independent variables in understanding human cognition and interactions.

These insights highlight the tremendous impact grammar has on emotions, behaviors, and perceptions across cultures as well as between individuals (Boyd & Richerson, 2005, p. 206). Semantics directly affects most aspects of human experience, including cultural belief systems, cognition, emotions, behaviors, evaluations, perceptions, affiliations, pair bonding, bonding mechanisms, social interactions, and even aggressive behavior. It seems reasonable and appropriate to account for this by constructing more *naturalistic* research designs for investigation of human brain function and behavior (Benjafield, 2007, p. 32; Grafman, 2002, p. 293; Delgado, 2007, p. 64; Giesbrecht et al., 2006, p. 104). This will require integration among many fields of science (Roepstorff, 2004, p. 1115; Norris & Cacioppo, 2007, p. 85; Edelman, 1992, p. 252).

Integration of Semantics and Grammar, Biases, and Cognition

Naturalistic approaches seem especially suited to studies assessing complex influences such as the role of higher-order brain functioning on beliefs (O'Doherty et al., 2007, p. 46) and ultimately, on accurate cognition and behavior. Humans have a highly developed frontal lobe system that, along with semantics and grammar, allows for higher-level executive functioning (LeDoux, 2002, p. 197). Even though cognition relies on measured integration of lower-level limbic and automatic cognition (Tranel, 2002, p. 351), the higher-level executive function and higher-order working memory are in a position to have the *last word*. Higher level and higher order as used here refer to executive functioning (DLPFC) with capacity for flexibility, explicit *objective* evaluations and abstractions, as well as considered thought and decision making. We can describe many higher cognitive processes as symbolic processes, including memory, attention, imagery, ideation, concept formation, generalization, abstraction, problem solving, thinking, reasoning, and planning (Logothetis, 2004, p. 849). Lower level and lower order as used here denote more rigid, automatic, stimulus-driven, or subjective emotional limbic appraisals and *subjective* implicit automatic cognition (Benjafield, 2007, p. 44-5, 51, 264-7, 307-9; Frith, et al., 2004, p. 265). This nomenclature allows for parsing of higher level and lower level brain function relative to grammar and cognitive accuracy (see Table 1). In human primates, language supports the option of making decisions based on reason rather than emotions. The last—and possibly the most influential—step in cognitive processing uses information abstracted from our personal history (Stuss et al.,

2001, p. 102). This information embodies the relative values of our personal cultural belief system that ultimately biases our choices and their consequences.

From culture to culture, semantic structure, grammar, and the use of language vary. The potential for higher-order abstractions appears enhanced in sophisticated cultures, and this potential seems associated with more highly developed language and advanced education. The processing of complex information gives rise to abstractions mediated by language (Luria, 1981, pp. 26-30). Abstractions range along a cognitive gradient from subjective to objective, depending on the degree of rationality of the grammar and processing. Across cultures, semantic, grammatical, and linguistic structural gradients range from dichotomous to multivariate, concrete to abstract, simple to complex, and few words and concepts to many. Language that supports more complex and abstract thought, at least in theory, seems to impart a cognitive advantage to *Homo sapiens*, "the human that knows they know" (Risberg, 2006, p. 3). By supporting higher-level abstract abilities, language provides the tools with which to think about how we think. Increased use of these higher-level abstract expressions of "thinking about thinking" apparently evolved fairly recently, around the sixteenth century or after (Benjafield, 2007, p. 235). This in turn allows us to discover the more objective, abstract principles of science that can lead to a more accurate understanding of the natural environment, human thought, emotions, and behavior (LeDoux, 2002, p. 176). This scientific understanding then forms the basis for the generation of a cascade of more accurate information for logical problem solving and appropriate objective decision making. Science offers a system for locating knowledge along a subjective-objective spectrum of classification. This distinguishes the subjective knowledge of cultural beliefs, what we *assume* we know, from objective scientific knowledge, what we can *demonstrate* we know.

When it comes to understanding humans and the human brain, it also seems important to consider the significance that semantics, grammar, and the resulting *beliefs* and biases have for human evaluations and interactions (Jones, 1998, p.14-23). *Brain Research* lags in consideration of the subjective cultural bias of semantics and language, and it rarely accounts for the relative semantic and grammatical accuracy of cultural belief systems and their relationship to thought, emotion, and behavior. Perhaps the relationship often goes unnoticed because we implicitly take our own belief systems so much for granted. The automatic semantics and grammar of our cultural belief systems lie at the center of our oldest and most strongly registered memory traces. They mostly operate implicitly—their day-to-day causal effects generally invisible to us—and we often take them at face value, without notice or challenge.

Table 1. Inaccurate versus Accurate Bias

Lower Level Inaccurate Irrational Bias	Higher Level Accurate Rational Bias
Faulty rigid assumptions; dogmatic beliefs, unsupported by facts, but stated as unquestionable "truths of the Universe" with questioning prohibited, "superstitious" ritualistic thought and behavior that promotes mind-brain dualism, subjective abstractions	Rational flexible assumptions stated as theories; hypotheses and conclusions supported by evidence, scientific testing, and mandatory questioning, "scientific" adaptable thought and behavior, mind = brain = mind/brain, objective abstractions
Rigid, maladaptive, with lower-level subjective bias, vertical subordinate communication	Flexible, adaptive, with higher-level objective bias, collateral communication
Absolute, static bias: certain, *"determinate,"* guaranteed	Variable, dynamic bias: uncertain, *"probability,"* not guaranteed
Cognition using dichotomous grammar limits freedom of executive function	Cognition using multivariate grammar expands freedom of executive function
Veridical bias: true and false, either-or, absolute, concrete, black and white; constrictive and restrictive, *not contextual*	Associative bias: abstract, gray, gradated; expansive and extensive, *contextual*
Predetermined certainty, all knowing, resulting in decreased frontal lobe requirements: *"afrontal"*	Relative uncertainty, inquisitive, resulting in increased frontal lobe requirements: *"frontal"*
Parental, demanding; adversarial	Adult, requesting; cooperative
Semantic inaccuracy: vague, poorly defined, with overgeneralizations: always, never, every, all, none, etc.	Semantic accuracy: specific, best definition and word use: frequently, infrequently, many, some, few, etc.
Rigid, implies no other choices: I should, I must, I have to, I need to, and I have got to. *"I am obligated"*	Flexible, implies choices; preferential: I prefer, I would rather, I would like to, I choose to. *"It is a choice"*
Tends to ignore inaccuracies of information, of thought process, and of event-level orientation; retroactive, "reactive"	Tends to promote accuracies of information, of thought process, and of event-level orientation; forward-thinking, proactive, "considerate"
Inaccuracies and faulty assumptions promote faulty and inaccurate cause-and-effect conclusions	Accuracies and rational assumptions promote more plausible and more accurate cause-and-effect conclusions
General unawareness of irrational cognitive process "Cultural belief system anosognosia"	General awareness of rational cognitive process and "Cultural belief system awareness"

Indeed, we simply and spontaneously tend to accept the grammar and beliefs of our culture as *factual* (Benjafield, 2007, p. 400). Unexamined inaccurate cultural beliefs, however, directly contribute to the mechanical use of inaccurate information along with rigid, dichotomous inaccurate cognition (Hooker & Knight, 2006, p. 317).

Evolution of Grammar

Human grammar contains many hidden, habitual attributes that directly affect our cognition, but we use them on a daily basis without inspection or consideration of reliability or accuracy. These hidden features include primitive, unscientific, subjective, and over-generalized object classifications, categorizations, groupings, and labeling. A relative comparison of language "grammar" equivalents between mammals, other primates, and human primates shows many similarities in basic object-action cognitive processing, demonstrating the high level of evolutionary conservation in mammalian brains. Brain areas for object-action cognition tend to function similarly in mammals and other primates, providing representations of stimulus pattern characteristics including object identification, object spatial location, and limbic mediated object risk-reward contingency value including evaluating intention of animate objects. However, human primates possess large visual and auditory association areas that represent recent evolutionary development for specialized vision and language processing networks. Theories of innate universal grammar postulated in linguistics may have derived from observed pattern recognition functions in the object-action-intention predisposition in mammals, primates, and human primates to the relative evolutionary conservation of brain function (Hauser, Chomsky & Fitch, 2002). However, a usage-based approach used here seems to offer a simpler model with more ecological validity for cross-cultural grammatical appraisals (Tomasello, 2004, pp. 642-5); Luria, 1981, p. 6), especially relative to evaluating the functional differences between multivariate and dichotomous grammar.

Language incorporates the skills of speech and grammar to provide a basic substrate for higher-level object and action description and identification, cognitive and emotional processing, and communication. Grammar also incorporates other sensory components, adding further descriptive value to stimulus object-action perception. Other primates have rudimentary language components but lack the sophistication of human syntax and grammar leaving their communication literal and mostly inflexible by comparison (Mesulam, 2002, p. 19). Mammals including other primates rely on sensory information to regulate environmental homeostasis, while limbic components provide valuable subjective risk reward contingency, information for actions. Homeostasis relies on feedback to maximize adaptation over time. This feedback may largely depend on semantics and grammar in humans. We have the benefit of words and grammar to assist in evaluating object salience and intention, contingency value, as well as action choice and planning. Contingency implies prediction. Prediction of causality requires awareness of the contingency between the intention and the action and between the action and its consequences (Portas et al., 2004, p.288). The accuracy of this awareness would seem pertinent for obtaining the most objective evaluations. Rigid limbic-driven imperatives dominate most other animals (Mesulam, 2002, p. 19). It seems that in humans, even though we have the added flexibility of our more sophisticated brain, we often simply augment or replace the limbic imperatives with rigid subjective semantic imperatives, especially regarding our subjective inferences and perception of the intentions of others. Humans have the option of lower level subjective limbic evaluations and higher-level objective evaluations not afforded to other mammals.

Vision and visual processing occupies a large part of the human brain. The addition of grammar creates value by providing a mechanism for parsing the categorization of environmental stimuli into discrete categories of objects, groups of objects, intentions, and actions (Nowak, 2006, p. 251-2. Vision plus grammar maximizes environmental object detection, subsequent labeling, and identification. Auditory detection and processing provides value, especially regarding the listening component of communication. The addition of grammar to motor function, the evolution of speech, and later writing, provided important components of communication. Symbol use and speech probably evolved around 50,000 to 100,000 years ago, while writing most likely evolved within the past 4,000 years (Striedter, 2005, p. 312-13). It seems to follow that early in language evolution, oral communication provided the major component of cultural transmission of information. Since this predates writing, story telling by narratives, influenced by implicit and explicit recollections and emotions, most likely provided the medium for informing each new cultural generation (Siegel, 1999, p. 333). In many cases, cultural belief systems, anecdotes, magical interpretations, myths, and story telling continue to trump scientific facts and statistical probability. This might also explain the prevalent human propensity to rely on primitive dichotomous grammar constructs and imprecise pattern recognition principles supported by superstitions, primitive rituals, over-generalizations, faulty subjective information, and faulty cause-effect conclusions (Vygotsky & Luria, 1993, pp. 138-9).

Cognition, an important piece of the human brain function puzzle, in large part relies on grammar. Evolutionary theory suggests that we benefit most as a species by achieving and maintaining consistent homeostasis with our dynamic environment. Furthermore, we can assume that we achieve the most effective object categorization and recognition, which guides our subsequent actions, by integrating and processing information pertinent to our event-level relationship with our environment. From this, we can conjecture that human long-term overall survival and success might depend on our ability to obtain the most accurate identification of objects and stimuli and the most accurate internal information about object relationships, and to process this information with the most accurate processing rules we can develop. To the extent that we can do this, we learn to make the most accurate assessments and devise the most reasonable methods for integrating variables involved in problem solving, for developing reasonable strategies, and for choosing the best actions to achieve a desired goal.

Evolution of Subjective Cultural Classifications

We might assume that language and grammar have evolved over evolutionary history as an adaptation for maximizing homeostasis of *Homo sapiens* with its environment, which includes other *Homo sapiens*. The phylogeny and ontogeny of words and grammars across cultures and history appear to offer some clues to the evolution of this adaptation. Words as linguistic symbols adapted by humans have evolved along side the social culture, as the defining language of the culture and the concomitant cultural belief system. The conventional usage learned by children includes the words and formulation forebears found useful in the past. Grammars evolve over many generations as our ancestors developed words, meanings, and grammar structure in response to changing environmental factors over time. This evolution developed from concrete utterances and symbols operating through cultural historical processes, rather than biological ones. Grammars passed along to subsequent generations, succeed as a product of familiarity, exposure, and the potential practical value they afford each new generation (Tomasello, 2005, p.13).

We might expect that the higher-level skills of more accurately identifying, manipulating, and interfacing with objects and other *Homo sapiens* in our environment would provide humans with survival benefit. Language offers a socialization advantage unique to humans, enhancing our ability to communicate, share information, plan actions, and understand the intentions of others (Risberg, 2006, p. 10). Nearly all language provides nouns for referentially labeling objects, and verbs for predicatively labeling actions or behaviors. It seems to follow that the more accurately we describe objects and predict their likely actions, the greater our ability for conceptualization of abstract representations. This improves our decision–making abilities, which in turn confers a greater adaptive advantage. Complex languages that enable much more objective higher-level abstract formulations would seem to offer an even greater adaptive value. However, as grammars become more complex and abstract, they can become less precise, with a potential for increased variance and errors. The importance of error monitoring, detection, and correction then increases. This suggests that assessments of abstractions based on accuracy might offer more precision and fewer errors, i.e., more objective, scientific abstractions versus more subjective abstractions.

Language evolved from primitive nonliterate cultures dominated by uninformed abstractions and subjective ways of knowing, and language still appears to convey some of that imprecision in cultures that continue to cling to subjective beliefs as facts. Science represents the other end of the knowledge spectrum, as demonstrated by the acceleration of more objective knowledge, i.e. scientific facts and technological advancements. The continuing dynamic seems to highlight the opportunity for rational intervention and the establishment of a more accurate reference point for measuring the objective descriptive precision of grammar, semantics, and abstract conceptualizations. Many of the abstractions humans use on a day-to-day basis fall into the category of *subjective abstractions* with inherent inaccuracies and irrational biases. *Objective abstractions*, on the other hand, use scientific principles and critical thinking skills to maximize accuracy, resulting in a more rational bias. This distinction, between cultural subjective categories and processing versus scientific objective categories and processing, offers a potentially beneficial reference point for human evaluations.

Ontogeny and Phylogeny of Grammar

How could humans evolve such a powerful dynamic evaluative capability as the human modern brain and continue to operate largely on primitive subjective evaluations? Infants begin their interactions with the world with generalized expressions, gestures, and utterances including cries, grunts, and coos. These and the concomitant facial expressions represent the primary form of communication with caregivers. Over time, infants begin to add single words and abbreviated short word phrases, as they master the naming of objects and actions, leading eventually to predication.

Verb development usually follows noun development. Labeling objects with names appears to help the child master the environment, and children seem biased toward using any new word as an object name (Bloom, 2002, pp. 92-9). After the child learns to use verbs and later to construct whole phrases, this functional vocabulary provides the foundation for the exponential growth for grammar development. Even in nonliterate cultures, the selection of objects to label reflects the linguistic ideology of the culture. Children apparently learn most of their early vocabulary simply by listening to the conversations around them. This applies particularly to preliterate children and to older children and adults in nonliterate societies (Bloom, 2002, pp. 118-9, 192). Much of this early vocabulary represents the building blocks of the background cultural knowledge and categories that structure and determine word meanings for abstractions related to space, time, causality, objects, intention, and possession (Tomasello, 2006, p. 54). This cultural knowledge also appears to bias the cultural frame of reference used in various other representations (Bloom, 2002, p. 247).

Children deduce early in language development that when adults refer to objects, they do so in terms of whole objects (Bloom, 2002, p. 92), i.e. they tend to over-generalize. A major part of human linguistic competence involves mastering by rote many routine formulas, fixed and semi-fixed expressions, idioms, and frozen collocations that objectively have somewhat unpredictable or inconsistent meanings (Tomasello, 2006, pp. 101-2). In many languages, including English, these expressions represent templates that, when used in conjunction with a coupler, can generate an almost infinite number of culturally *subjective* abstractions. A common grammar component across many languages is the copula, or coupler. In English, the verb "to be" represents the primary copula, and its use appears early in development (Tomasello, 2006, p. 255). Unfortunately, despite its ubiquity, "to be" is generally insensitive to pattern specificity, time specificity, and context specificity, which makes it very useful for subjectively over-generalizing labels but at the cost of reduced precision and distorted perceptions.

The utility value of the predicate form of "to be" encourages subjective generalization of behaviors (actions) to objects. This allows indiscriminant objectification of dynamic processes by semantically converting any subjective condition into a seemingly objective noun (i.e., we can say, "He *is* a *failure*," instead of saying, "He *failed* at this particular effort" or "She *is bad*," rather than "She acted *badly* last night"). The verb "to be" represents great utility as a very handy general label-making tool. In light of this generalization utility, "to be" might represent the basis for primitive verb evolution. Cultural labels often represent subjective, discriminatory, rigidly held prejudices, biases, and habitually defended beliefs, supported in lieu of objective scientifically derived classifications or alternatives. Objectification, or subjective labeling of humans or groups of humans as objects, presents a biased, usually derogatory "image" of those humans. Human adults and infants reflect this bias (Bloom, 2002, pp. 93-4). Other evidence suggests that the mere presence of labels may encourage people to exaggerate differences between groups (Bloom, 2002, p. 254).

Cultural belief systems also rely on prescriptive, imperative, rigid, authoritarian, and sometimes intimidating auxiliary verbs, or "helping" verbs (should, must, have to, need to, ought to, got to, etc.) that do not convey contextual specificity. Descriptive verbs describe the reality of the world, while prescriptive verbs describe how the world *should be*, usually based on unspoken assumptions. Unfortunately, prescriptive verbs usually express a subordinate frame of reference bias held by the *should-er*, but not necessarily known or accepted by the *should-ee*. "Need" is often overused subjectively. Humans need food, water, shelter, and air to survive. We may want or desire other things but they hardly meet objective criteria as needs. Many of these prescriptive and imperative verbs represent arbitrary cultural artifacts that deftly sustain the cultural belief system and exclude rational choice or consideration of reasonable choice-outcome paradigms. Each decision has many choices and variable consequences. Prescriptive and imperative statements demand a predetermined *choice*, with insensitivity to awareness or consideration of the *context* or the nature of the decisions and the probable outcomes involved (Capaldi, 1987, p. 17). Prescriptivism supports dominance grammar and consequently supports and sustains cultural belief systems.

Factitive and causative verbs combine a direct object and a phrase to apply a certain characteristic or a change in status to an objective complement. One such verb, "to make," works hand in hand with prescriptive statements. "You *make* me sad, you *made* your father angry, you *made* me act that way." These subjective verbs enable an individual or group to transfer to others the responsibility for their own thoughts, emotions, behaviors and choice-consequence outcomes. Factitive and causative verbs probably made sense historically in the primitive evolution of grammar, language, culture, and learning. Perhaps due to their social utility, they survive even in well-educated cultures by providing a vehicle for maintaining cultural belief systems and superstitions, transferring responsibility, and blaming of others.

Humans also often apply faulty knowledge and confuse coincidence or correlation with cause and effect (Skinner, 1953, p.84-5). We learn these habits as children, by listening to the conversations of adults. Unfortunately, "children aren't like scientist who have theories; they are like scientist before they have theories, trying to make sense of some domain they know little about" (Bloom, 2002, pp. 168-9). Children physically grow into adult humans but generally retain their subjective cultural belief system biases. Science requires inferences based on scientific inquiry with objective observation, and correlations with statistical probability. "Science attempts to find out how things really are, not just how they appear to be." (Bloom, 2002, p. 169) Subjective cultural belief systems often result in this true or false, dichotomous grammatical underpinning to human cognition.

Dichotomous grammar, consisting of veridical, either-or, black or white structures, continues this pattern of culturally learned conventions. Culturally derived dichotomous cognition automatically preempts objective, multivariate cognitive strategies for evaluation, even in adult humans. The verb *to be*, imperative prescriptivism, and either-or cognition represent complementary candidates for the primary source of primitive dichotomous grammar evolution. Even though these habits become automatic and implicit in a sense, they can become apparent with ongoing effort and inspection. However, we have little chance of noticing them, finding them, or correcting them if we do not even think to look in the first place (Jones, 1998, p.33).

Evolution and Dominance Hierarchies

An evolutionary model seems to offer some plausible explanations for the relationships between culture and language. If we start with the assumptions that dominance hierarchies have historically supported the evolution of adaptability in human social systems, and that survival depends on group functioning, then it seems to follow that hierarchies controlled through reinforcement and punishment by dominant individuals or groups of individuals have historically conferred an adaptation for survival and reproduction. This applies to many other species as well as to humans. Indeed, in modern human interactions and language, we may still observe the slow evolution of these fundamental dominance features. If language and grammar have evolutionary characteristics, we might benefit from considering how these adaptations evolved.

A language, in some ways, resembles a species, as it defines social groups. Within most species of languages, there is variety, and most large groups exhibit significant differences in dialects and colloquialisms. Greater descriptive variety and complexity in language and grammar can enhance higher-level abstractions and flexibility; in contrast, constricted or subjective dichotomous homogeneity shifts the speaker towards concreteness and rigidity. If descriptive complexity enhances the potential of language features, including higher-level abstraction, and offers an adaptive value, we would expect to see languages with these features proliferating over time and becoming more dominant; indeed, this seems to be the case. Descriptive variety appears to endow language with this adaptive evolutionary trait of diversity, in much the same way that variety and diversity supports the natural selection processes described by Darwin in *The Origin of the Species*.

This spectrum of linguistic and grammatical variety, ranging from concreteness to abstraction, represents various gradients used for cognition: from static thinking to plastic, rigidity to flexibility, dichotomous to multivariate, resistance to diversity and change to enhancement of diversity and change. We observe a similar evolution in organisms from single-cell simple systems to multi-celled complex systems, with each level of complexity usually enabling enhanced flexibility and adaptive value. Cultures tend to parallel this progression from simple to complex, brought about and supported in large part by complex language and grammar. If we apply evolution theory to grammar and social development, we may assume that the *beliefs* that helped our ancestors and parents reach the age of reproduction and produce offspring may, in a sense, pass on to their children as a inherited adaptive value, however rigid or inaccurate. When contemplating how languages and cultural belief systems could possibly demonstrate such rigid resilience over time, it helps to remember the span of human evolution represents a very brief time evolutionarily. In our earliest stages as humans, we most likely benefited from a strong sense of confidence in the historical success and survival of the group. In this brief window of social evolution, in some ways we see relatively small transmutations from rigidity to flexibility, while our brains have progressed biologically to the point where they can support abundant flexibility. The firmly ingrained language of social rules and beliefs, supported by rigid grammar structure, seems to impose impediments to change.

Indeed, dominance hierarchies in humans—characteristic of our genetic inheritance and conditioned by our environment—have a large influence on this stodgy rigidity. Our brains contain hard-wired circuits for aggression, territoriality and competition for resources, dominance status, parental-familial-affiliation, defensiveness, and irritability (Wingfield et al., 2006, p. 180). These circuits represent a significant inertia and resistance to the evolution of more objective and rational language use that can replace our typical irrational, dominance-related expressions of aggression. From individuals to whole cultures, we see dominance and rigidity perpetuated by the grammatical language habits acquired in early development. Our learning mechanisms (including *modeling*, instrumental learning, and associative conditioning) largely depend on implicitly embedded language and grammar, which makes it difficult for us to discover and overcome the inertia of our belief systems. This learning, supported by semantics, forms the basic template of *what we believe* and *how we think* for most of our lives. Our language and grammatical processes provide the vehicle for culture, social structure, and adaptation that, theoretically and potentially, enable us to operate on a gradient from a subjective irrational bias to a more objective rational bias. Culturally, we tend toward the former, while cognitive accuracy offers a way to move closer to the latter.

Cultural Social Inheritance

Through the mechanisms of genetics and learned belief systems, our thought and behavioral patterns reflect our personal, social, and cultural histories. Genetically, we inherit a basic blueprint for how our brains operate. Culturally, we acquire a blueprint for what we think, how we think, and for expected behaviors. Inevitably, as humans we inherit cultural beliefs representing a long line of learned and rigidly held subjective inaccuracies that bias our perception. Becoming aware of the potential inaccuracy of what we know, *or think we know*, allows us to make corrections and to think more critically. We measure cognitive accuracy by the relative distance or gradient between the unexamined, inconsistent, and irrational yardsticks we have inherited and the validated, external, more reliable, rational reference points we have identified through science. The shorter this distance, the more objectively, rationally, and accurately we think.

Most people contend they think accurately, rationally, and logically. However, they generally base their contentions on their own usually unexamined, inaccurate, and irrational frame of reference. For the most part, our individually inherited grammar and cultural belief systems subjectively bias our thoughts and perceptions, even of ourselves. This bias becomes apparent when compared with a rational reference point or standard (see Table 1). Awareness of this irrational bias opens the door to the adoption of standards with more objective accuracy and reliability, resulting in the potential for a more objective rational bias.

HUMAN BRAIN MODEL

Neuroscientists sometimes describe normal human brain functioning in terms of a computer model (Avrutin, 2006, p. 49; Benjafield, 2007, p. 14-18; Braver & Ruge, 2006, p. 338-9; Neisser, 1967; Panksepp, 1998, p. 20; Shiffrin & Atkinson, 1969; Toates, 2007, p. 17; Baum, 2004, p. 7). Our brain (*hardware*) comes initially from our inherited genetic blueprint; environmental learning makes the major contribution to our stored memory information (data), and grammatical process information (*software*) (Schmalhofer & Perfetti, 2007, pp. 180-81). The brain stores memories as information in various storage areas, similar to a computer's hard drive. A portion of this storage contains grammatical templates for *processing* information—the rules of how we think, similar to computer software that determines how information is processed (Fuster, 2003, p. 55). Like computers, humans may acquire or develop faulty or pathological hardware, software, or data. Both computers and humans get the most accurate results with the most appropriate hardware, most up-to-date adaptive software, and as accurate and timely data as possible. This enables the most reliable choice-outcome conclusions at a desired point in time. Neither computers nor humans can produce satisfactory results with inaccurate, out-of-date data or faulty, inflexible software. We tend to build error detection into our computers, and to upgrade and install new versions of software as new information and technologies become available — it seems logical to look for ways to do the same with our own human brain computers. If instead we approach the brain as a black box with only input and output, we would most certainly have very much interest in the unknown processing going on inside the black box. Fortunately, we do have a growing knowledge of this processing in the human brain. The quality and precision of the output seems to correlate with the quality of our information and quality of its processing. With direct evidence of the adverse affects of irrational grammar usage on cognition, emotions, and behavior, we can abandon the ancient perception of the brain as a mysterious black box, and scientifically apply what we have learned about it to improve the way we make use of it.

Frontal Lobe Integration: Executive Functioning and Working Memory

The higher-level executive working memory (Baddeley, 2002, p. 246) of the brain's frontal lobes (specifically, the integrated functioning of the dorsolateral prefrontal cortex (DLPFC), along with frontopolar cortex (FPC), Broca's area, temporal, temporoparietal and association areas, etc.) compares to a computer's random access memory (Mesulam, 2002, p. 26). The information the brain uses to make decisions with compares to the data stored in the computer. The DLPFC and FPC appear to play an important role in the integration of internal and external appraisals to adapt to changing conditions (Gazzaley & D'Esposito, 2007, p. 188; Risberg, 2006, p. 6; Wagner et al., 2004, p. 714) and with explicit empathy with others (Decety, 2007, p. 258-60; Ferstl, 2007, pp. 87-90). This dynamic integration of information enables critical error monitoring, error detection, and error correction. The left DLPFC plays a large role regulating language function and sequencing while the right DLPFC has more involvement with the processing of interpretations, inferences, and concepts, awareness of novelty, situational and emotional appraisals, as well as autobiographical memory (Schmalhofer & Perfetti, 2007, p. 184; Long et al., 2007, p.330; Tapiero & Fillon, 2007, p. 365; Siegel, 1999, p. 331). The OFC assists with a parallel role in mediation of emotion and object-affect associations (Mega et al., 2001, p. 23).

The anterior cingulate cortex (ACC) appears to play a pivotal role in motivation-related error detection (Niv, 2007, p. 369), novelty/complexity detection and performance monitoring (Braver & Ruge, 2006, p. 322-24) by interfacing with the DLPFC and limbic, temporal lobe, parahippocampal gyrus, and automatic subcortical association pathways (Kaufer, 2007, p. 49-52). When automatic pathways offer incongruent motivational resolutions to homeostasis, the DLPFC generates congruency. Over time, "congruence strategy" efficiency will benefit from up-to-date information, continual monitoring and improvement of error detection strategies and increased error resolution precision. Error correction may be enhanced deliberately and rationally, by updating the accuracy of perception i.e. information, process, and event-level accuracy, or reflexively and irrationally, by explaining away discrepancies and preserving subjective beliefs and perceptions despite their inherent inaccuracies. Without frequent evaluation of the relationship between our behavior and the environment relative to reality, we can easily fall prey to stimulus bound reactive behaviors (Luria, 1966/1980). As a part of the broader network involved in adaptive decision-making (Lee & Seo, 2007, p. 108), the DLPFC can conduct more objective, considered, and deliberate routine monitoring (Goldman-Rakic, 1995) of appropriate cognition, emotion, and behavior (Braver & Ruge, 2006, p. 321), perhaps enhanced by operating with a perspective biased toward accuracy and objectivity. This objective accuracy bias potentially offers improved error detection and correction with enhanced risk prediction (Preuschoff & Bossaerts, 2007, pp. 142-45).

Even though the orbitofrontal cortex (OFC), limbic system, hippocampus, striatum, etc., play a critical and often parallel role in working memory and decisions (Wagner et al., 2004, p. 709), "there is a hierarchy" (D'Esposito & Postle, 2002, p. 177; Curtis & D'Esposito, 2006, p. 295, fig. 9.8). "Executive functioning requires the integration of prefrontal and subcortical activity", (Mega & Cummings, 2005, p. 18). Semantics, grammar, and language seem to contribute, directly or indirectly, to accurate cognition at each level. A series of parallel networks, including the subcortical cognitive pathways, provide information for executive decisions and explicit and implicit automatic cognition, allowing for adaptive interactions with environment (Chow & Cummings, 2007; Salloway & Blitz, 2002; Lichter & Cummings, 2001, Middleton & Strick, 2001). This hierarchy allows for top-down use and regulation of subcortical systems to enhance monitoring, and correction of contingent input-output incongruence (Stuss, et al., 2001, p. 101). An anatomical and functional viewpoint of these networks seems to offer some relevance for understanding this hierarchy.

The frontal lobes comprise two basic anatomical and functional systems (Pandya & Barnes 1987), the dorsal system consisting of dorsolateral and medial portions of the frontal lobes, and a ventral system consisting of the orbital portions of the frontal lobes. The dorsal system has interconnections with the posterior parietal lobes and cingulate gyrus, and deals with sequential processing of sensory, spatial, and motivational appraisals of external environmental objects and stimuli (the *where* of the event). The ventral system has interconnections with limbic networks involved in regulation of internal homeostasis and emotional conjectures about perceived external objects (the *what* and *why*, or intention), as well as appraisals of stimuli salience, valence, and motivational relevance (Ogar & Gorno-Tempini, 2007, p. 59). The two basic divisions along with their two primary subdivisions are:

- DLPFC, providing executive functions of deliberate, explicit executive cognition/explicit working memory involved in processing higher-level cognition. The superior medial prefrontal cortex (a.k.a. ACC) functions as an automatic explicit/implicit integrative center for cognitive-behavioral (attention-motivation and possibly error monitoring) and emotional-autonomic-motor neural networks (Chow & Cummings, 2007, p. 25; Kaufer, 2007, p. 49).
- OFC, providing limbic regulation and ACC interaction involved in emotional integration of automatic, explicit/implicit, emotional working memory (medial OFC), including ventral striatal, medial temporal lobe (MTL), amygdala, hippocampus, and hypothalamus. In addition, the lateral OFC is involved in social working memory processing for emotional components of social motivation and

cognition (Chow & Cummings, 2007, p. 30-1; Salloway & Blitz, 2002, p.10; Mesulam, 1985).

Each of these systems may contain subsystems along with accessory parallel frontal-subcortical circuits (FSC), consisting of cortico-striato-pallido-thalamic-cortical loops. There have been at least five to seven proposed general loops including skeletomotor, oculomotor, dorsolateral prefrontal, anterior cingulate, lateral orbitofrontal, medial orbitofrontal, and possibly inferotemporal/posterior parietal (Alexander et al., 1986; Middleton & Strick, 2001). Each of these loops appears to be comprised of multiple parallel segregated circuits. These pathways offer multiple levels of parallel processing of information for cognition, emotions, and behavior, including extrapyramidal, motor, and speech systems probably including rule-based cognitive processing (Poldrack & Willingham, 2006, p. 130-1). The basal ganglia have direct input from the large association cortex in the parietal, temporal, medial, and frontal areas, as well as the hippocampal formation and the amygdala. There is also large outflow directed towards the frontal cortex via synaptic links in the thalamus, outflow to temporal and parietal lobes linked to executive functioning (Graybiel & Saka, 2004, p. 495-7) and open pathways to inferotemporal and posterior parietal areas.

In addition, other pathways provide integrated reflexive, or innate, lower-level automatic, stimulus-response pathways, responding to potentially threatening stimuli, i.e. loud noise, fast movement, gestures, as well as to threatening facial expressions and involvement in reflexive empathy. The frontal lobes evolved to manage the pyramidal pathway and to implement cognitive decisions in adaptation to a dynamic environment by incorporating a wealth of sensory input, using higher cognitive function to prioritize that input, and choosing the most effective response, by ongoing evaluation of shifting priorities, initiating action, and monitoring its execution (Chow & Cummings, 2007, p. 38).

The separate but parallel dorsal and ventral anatomical systems appear to represent two functionally distinct hierarchal networks relative to objective higher level DLPFC and subjective lower level OFC cognition respectively. "'Executive cognitive functions' are defined as (and the term is limited to) high-level cognitive functions, believed to be mediated primarily by the LPFC (lateral prefrontal cortex), that are involved in the control and direction (e.g., planning, monitoring, energizing, switching, inhibiting) of lower level, more automatic functions" (Stuss, 2007, p. 293). Executive functions include the ability to solve a complex problem and organize a volitional behavioral response in a temporally informed manner. They also encompass the learning of new information, the systematic searching of memory, activation of remote memories, appropriate prioritization of external stimuli, attention, generation of motor programs, metacognition, and probably the use of verbal skills to guide behavior (Chow & Cummings, 2007, p. 29; Cummings & Miller, 2007, p. 15). Language, semantics, and grammar usage associated with these distributed neural networks may explain the integration of orbitofrontal/emotional and dorsolateral/cognitive processes and resolve the long-standing difficulties in the reconciliation of internal integration and processing of cognition and emotions, especially for higher order processes such as decision making and social perception (Gazzaley & D'Esposito, 2007, p. 190-1).

The DLPFC modulates cognition and human environmental interaction from the top down via integration of externally driven information with internally represented information. The relative effectiveness of cognition seems to depend on the quality of external information received or perceived, and on the perceptual quality of information stored in long-term memory, including "beliefs" related directly to information, information processing, emotions, responsibility, and intentional perceptions of others (Gazzaley & D'Esposito, 2007, p. 192-7). Language may directly mediate the quality of cognitive integration as measured by objective accuracy. Executive functioning depends on the integrity of instrumental functions such as language (Cummings & Miller, 2007, p. 16). Semantics and grammar appear to act as the common denominator between higher-level executive functional networks and lower-level limbic system functional networks, supporting appraisals for the identification and perception of object/stimulus salience, emotional valence, actions, and reactions.

Human evolution seems to have "thoughtfully" overlaid semantic representations onto our cognition. Words and concepts bind to sensory information, creating a kind of a sixth sense—specifically, a semantic sense (LeDoux, 2002, p. 203). This allows language and grammar to provide a substrate for human cognition using words to assign values to stimuli, emotions, behaviors, perceptions, belief systems, etc. Culture provides the foundation for the development of values by grounding specific cultural beliefs in grammatical semantic usage, coupling words and beliefs with historically learned reward contingencies during development. Words have become a sense for detecting and coding predictive relationships among stimuli, objects, responses, and rewards. As such, words assign the relative emotional and motivational significance for the value of rewards and the anticipation of rewards (Phelps & LaBar, 2006, p. 432). These evaluations are automatically mediated by the OFC (Petrides & Pandya, 2002, p.45-46) and depending on task demands, will often default to lower level automatic grammar habits managed by the striatal complex (Poldrack & Willingham, 2006, p. 134-40). Cultural beliefs thus underpin the fabric of human social interaction, by relying on learned, automatic implicit grammars that allow speech to flow effortlessly. Mostly learned by implicit inductive inference in childhood (Nowak, 2006, p. 263), habitual grammar forms a silent, "ritualized" soundtrack for our adult cognition.

Words and grammar give us a unique descriptive ability to encode environmental cues with the potential to maximize efficient usage of neocortical associations. Depending on task demands and stimuli, association mechanisms generally run implicitly and almost continuously in the background, evaluating object features and place associations, risk-reward associations, and past related choice-outcome associations. As complexity, novelty and unfamiliarity increase (Anderson et al., 2002, p. 505), and when higher-level abstract demands increase, the brain typically achieves more preferred outcomes by using the executive functions of the DLPFC and FPC (Curtis & D'Esposito, 2006, p. 293-95; Braver & Ruge, 2006, p. 330). However, in routine situations, lower level limbic and habitual cognition often suffices.

The lower level OFC provides limbic regulation and valuable utility calculations (Glimcher et al., 2005) while monitoring somatosensory information, emotional valence, relative risk-reward contingencies, and social salience. It also contributes to inhibition of inappropriate responses (Rolls, 2002, p. 370). Except for the sophisticated language system, the human OFC somewhat resembles that of other primates. However, in humans, the OFC also automatically provides ongoing bottom-up regulation of the limbic system using the default semantics, language, and grammar from our cultural belief system, thus reinforcing the subjective social, neuroeconomic, and relative psychological values that define our cultural bias (Kahneman et al., 1982; Padoa-Schioppa & Assad, 2006). When subjective cultural belief system grammars are edited to operate more accurately (i.e., yielding a more scientific belief system), the bias shifts towards the more objective end of the value probability continuum, operating closer to the rational-cognitive end of the evaluation scale rather than the emotional-motivational end (Schultz & Tremblay, 2006, p. 195).

Information Content and Information Processing

The top-down executive functioning of the DLPFC likewise relies heavily on semantics, language, and grammar to perform critical integration of choice-outcome determinations and inhibition of inappropriate responses. The lateral prefrontal cortex enables greater complex abilities and flexibility in human and primate behavior (Striedter, 2005, p. 309). The newer higher level DLPFC adds the potential for more accurate rational thought onto the older lower level limbic brain by enabling humans to reason (Panksepp, 1998, p. 20). Efficient prefrontal cognition relies on the quality of information received from other cerebral regions (Anderson et al., 2002, p. 506). While rigid cognitive inaccuracies can impede the discovery and executive execution of the most logical and reasonable decisions, shifting toward an accurate and rational adaptive bias enhances objective cognition, enabling more reasonable choices and outcomes, emotions and behaviors. In terms of the somewhat simplistic computer analogy, the DLPFC executive function, probably aided by information integration involving the FPC, has the capacity for objective appraisal, reappraisal and flexible editing, while the OFC is limited to subjective emotional-based read-write operations, or subjective contingency appraisal and reappraisal. Evidence indicates that task relevant semantic knowledge develops in a relatively automatic (bottom-up) fashion or in a more controlled (top-down) manner (Wagner, et al., 2004, p. 715) probably flexibly regulated by the DLPFC. The DLPFC sits at the top of the PFC as the captain of the ship, demonstrating a higher level of competency when provided with the resources of the most accurate and the most up-to-date information.

Even though the limbic working memory can adjust responses, it does so primarily based on reward contingencies, reinforcement, and emotional valence spectra (Frith et al., 2004, p. 265). The OFC adds some subjective flexibility, but our ability to objectively change and integrate the accuracy of *what we know* and *how we think* relative to the environment depends most heavily on the reappraisal and edit functions of the executive DLPFC working memory. The DLPFC compares internal stored memories and habits with external, objective, up-to-date information, with the rational capacity to temper our cognition, emotional responses, and automatic behavioral and grammatical responses to stimuli (Mesulam, 2000, p. 93). The DLPFC has the potential to edit our culturally inherited grammar and provide improved cognitive accuracy for error detection and correction. In addition, the FPC (BA 10) appears to have strong connections with temporal areas that support auditory working memory (Barbas, 2006, p.51; Davachi et al., 2004, p. 670).It appears that, by holding information online during secondary task manipulation in working memory, the FPC supports DLPFC edit functions (Petrides, 2005).

The DLPFC appears to have the ability to directly select or reject individual strategies and may possibly edit grammar through direct mechanisms. However, the edit function most likely relies on explicit DLPFC-initiated reentrant "working memory" rehearsal loops for corrections (Curtis & D'Esposito, 2006, p. 283). These loops develop by practice according to Hebbian rules (Hebb, 1949), and the information at some point becomes routine and then automatically provided "preferentially" by cortical association areas (Braver & Ruge, 2006, p. 328).

Corrections might also occur over time due to environmental contingencies, modeling, and instrumental and conditioned learning. Cognitive therapies take advantage of this edit function by using explicit rehearsal to teach patients more advanced, and more accurate, cognitive strategies. This provides increased cognitive accuracy for making more rational decisions and adaptations in a variety of situations. The overall efficacy of this approach probably accounts for the increasing success and acceptance of rational emotive and cognitive behavioral therapies (Ellis & Harper, 1997, pp. vii-xiv; Beck, 1976, p. 4; Wright, 2004, pp. xv-xx). The corrective effect of the edit function on cognitive accuracy greatly enhances the probability that the higher-level objective executive functioning of the DLPFC will have the most rational last word in decision-making (D'Esposito & Postle, 2002, p. 177).

The frontal lobes use stored processing rules for executive functioning, evaluating and prioritizing the available information to determine the best choices for reaching our goals. Declarative memory (i.e., information about how the world works, both *subjectively* and *objectively*) generally appears to be processed for storage by limbic system components, including the hippocampus (Duvernoy, 2005, p. 28), which has significant connectivity with the amygdala (Mesulam, 2000, pp. 2, 58-63). The hippocampus assists in encoding memories with place preference and the amygdala with the assignment of emotional valence (LeDoux, 1996). This information includes semantic and grammatical memory, episodic memory, verbal memory, and biographical memory, along with the rigid rules of our cultural belief systems. Retrieved memories tend to return with the emotions, or limbic valences, that accompanied their initial storage (Phelps & LaBar, 2006, p. 430), albeit modified to some extent by our present emotional state and by time-related decay. This allows past learning to condition present experience and explains to some extent how our cultural belief systems rule us so tenaciously and with such strong emotions. Fortunately, we can ameliorate these subjective emotional reactions by evaluating and editing inaccurate grammar and cultural beliefs, including our faulty assumptions about the cause-effect relationship between thoughts and feelings (Ochsner, 2007, p. 107; 2005, p. 253).

Cognitive Processing

Words and grammar enable us to describe environmental cues, making maximum use of neocortical associations. The level of complexity relates to increased working memory demands. The ability of the frontal lobes to use working memory optimally depends heavily on the availability and quality of the objective process information received from other cerebral regions (Anderson et al., 2002, p. 505) as well as accurate environmental perception. The frontal lobes function best with accurate and timely up-to-date information combined with accurate thought processes i.e., accurate data and the most appropriate, flexible software (i.e., how we think). Thinking uses process information for the assessment and integration of our internal learned information with external environmental information in order to regulate our interaction with the environment and maintain a probabilistic model of anticipated events (Raichle, 2006, p. 13). The software enables the frontal lobes to process executive decisions and to help regulate our emotions appropriately and provide overall homeostasis (Malloy & Richardson, 2001, p. 128).

Grammatical process memory (i.e., acquired and developed rules of thinking) directly biases not only how we process internally stored information but also how we perceive environmental information, cues and stimuli (Decety, 2007, p. 284). Learned, rigid, *dichotomous* process rules preempt flexible thought processes and distort the accuracy of object identification and evaluation. As we might expect, *rigid*, inaccurate, irrational process information leads to inaccurate, irrational information processing, and this in turn yields inaccurately biased executive decisions. Faulty, inaccurate, rigid, dichotomous software bias tends to yield subjectively biased, inaccurate choices and outcomes, which thwart the *flexibility* offered by higher-order, multivariate executive functioning—completing the cycle by reinforcing rigid cultural subjectivity. "No complex system can succeed without an effective executive mechanism, 'frontal lobes.' But the frontal lobes operate best as part of a highly distributed, interactive structure with much autonomy and many degrees of freedom" (Goldberg, 2002, p. 230).

HUMAN BRAIN AND COGNITIVE DEVELOPMENT

Where do inaccurate, irrational thought processes come from? How could we have learned our culturally inherited, faulty, inaccurate, irrational thinking without realizing it? As humans evolved higher-level brain structures, they underwent biological changes that freed them from the constraints of simplistic, concrete language processes and offered the potential for complex, abstract, associative reasoning, made possible by the dramatic evolution of the neocortex and frontal lobes. The capacity for improved frontal lobe reasoning evolved in humans alongside our development of semantics, grammar, and language skills. This same evolution from concrete thinking to the potential for abstraction occurs in an individual's frontal lobe development and connectivity as they grow from childhood to adulthood. Concrete thought processes evolve into more flexible, associative, abstract cognitive capabilities. During these developmental stages—towards puberty and into young adulthood—the frontal lobes continue to mature, with increased connectivity and myelination, especially in the DLPFC (Dennis, 2006, p. 135). This development—along with language, learning, and education—increases the potential maturity of our decision-making skills (i.e., executive functioning) and heightens our ability for cognitive awareness. The frontal lobes play a critical role in this process (Stuss et al, 2001, p. 108).

An episodic increase in cortical pruning takes place in young adolescents around 11 to 14 years of age. This pruning appears to select out the least used or weakest neural connections to make room for the more focused information processing required for improved complex problem solving, social and sexual maturation. According to Hebbian principles (Hebb, 1949), the most used or strongest connections are preserved. Unfortunately, the strongest and most used connections in young humans tend represent the belief systems and associated rigid dichotomous grammatical processing information of their childhood culture, which may become even more firmly embedded in the memory storage of the young brain during this period. The timing of these changes may possibly serve to enhance the resiliency and persistence of these ingrained belief systems.

Cultural Inheritance

As we grow into adulthood, neural efficiency and connectivity to the frontal lobes increases, along with the increasing use of acquired information and grammatical processing rules to make executive decisions. We tend to use our acquired information as if it were factual (Jones, 1998, p. 46), automatically following rigid, dichotomous processing rules that operate outside our awareness. We erroneously tend to identify the first thoughts that come to mind as "facts" and then execute heuristic (arbitrary rules of thumb) justifications and prefabrications to explain away errors based on these "facts." We implicitly assume good-bad relationships between words and reward contingencies learned during our personal developmental history. This may possibly contribute to locking in our cultural belief systems at the expense of more effective, higher-level executive functioning. We seem to favor irrational automatic rules over the acquisition and application of more accurate and up-to-date information. Our decision-making process tends toward a subjective, bottom-up frame of reference with our own personal cultural bias, driven by information filtered through our subjective personal historical matrix, and distorted by the inevitable, implicit, grammatical inaccuracies embedded there.

Our cultural bias not only reflects the ineffective cognitive habits and grammatical inaccuracies of our ancestors, it also represents the rigid, culturally biased misperceptions we inadvertently inherit from generations of unscientific, misinformed, and most often poorly educated elders. Our underlying beliefs derive in large part from their ritualized concrete thinking, faulty assumptions, coincidences imagined as cause and effect, superstitions, myths, magical thinking, etc. (Benjafield, 2007, pp. 321, 333-4). Because children have little experience with which to evaluate different concepts, they uncritically absorb whatever they encounter without regard for its rational usefulness or accuracy. They implicitly and unconditionally absorb rules and information expressed as *truths* without the full benefit of mature executive functions. They tend not to question the factual basis of information, the logic of assumptions, or the reasonableness of conclusions, nor do they have the acquired grammatical framework to do so (LeDoux, 2002, p. 96). This results in a subjective discriminatory bias. In a sense, children are "imprinted," (Lorenz, K. 1965) blind to other possibilities (Benjafield, 2007, p. 267; Shilpa et al., 2007, p. 53).

Information and Process Bias

Concrete learning from early developmental stages tends to accumulate in memory-storage areas, along with the misinformation absorbed from cultural belief systems. The stored information usually reflects the dichotomous hierarchal processes of the parent-to-child interactions under which they formed. If their cultural belief systems remain unchallenged, adults usually exhibit parent-to-child *authoritarian* grammar characteristics in their thinking and interactions, with irrational, prescriptive and imperative demands such as *should, must, have to, need to*, etc. These parental, subjective, dichotomous demands typically "trump" multivariate choices and impede adults from taking responsibility for objective choice-consequence decisions. Even though the parent-to-child interaction style appears to have evolved to beneficially manage and control the child's behavior until the frontal lobes develop, it unfortunately results in implicitly carrying irrational, dichotomous, authoritarian, parental thought and speech habits into adulthood.

Uncertainty and Deviation from the Mean

This irrational thinking usually passes down through many generations and carries into adulthood in part because of our tendency to gravitate toward the familiar (Benjafield, 2007, p. 386) and away from the unfamiliar. Humans have a tendency to experience rewards, or gratification, for sticking with the familiar, and punishment, or anxiety, for venturing into the unfamiliar. From an evolutionary view, this makes a lot of sense, because the unfamiliar may represent a potential danger, while the familiar has at least theoretically been evaluated and found to be safe (Keller & Chasiotis, 2006, p. 277). Familiar events or those with outcomes we deem certain deviate very little from the mean, and as such, generally receive a positive valence with a higher *probability* of favorable outcome.

Uncertain events can deviate significantly from the mean and usually receive a negative valence and a lower favorable probability value. This tends to give certainty the upper hand over the perceived ambiguity and unfavorable probability of uncertainty. This creates a problem for humans determined to cling to illusions of certainty in an uncertain world. However, irrational thinking and heuristics, commonplace rules absorbed from the dominant culture, magically repair these deviations from reality. We use heuristics that rely on faulty grammar and faulty logic to whitewash our errors and inaccuracies and to temper the inherent increase in error variance. This allows humans to alter their perception of reality to explain away any variance by clinging to the irrational *certainty* of their faulty beliefs. Unfortunately, the erroneous irrational *perception* that variance has decreased cannot mitigate the *actual* increase in real error variance and outcome imprecision.

In a sense, the familiar represents comfort, and we human creatures favor our comfortable habits. Humans so favor the familiar that they will often tolerate a great deal of discomfort to hang on to it, sometimes enduring even catastrophe before they will consider the alternative of change, or deviation from the mean. Unfortunately, humans usually tend to err on the side of favoring the familiar and generally receive no logical framework or training to help us develop objective, reasonable, innovative methods for seeking rational alternatives. Indeed, many cultures actually punish attempts to promote cognitive accuracy, some very severely, because it challenges authority and cultural beliefs. This usually prejudices us toward choosing information based on habit, familiarity, and subordination rather than rational utility or applicability, regardless of what might represent our best interest from a more objective perspective.

Authoritarian Communication

Due to the parent-child environment in which they learned these concepts, most adults implicitly continue to use the familiar parental cognitive process in their adult interactions. This hierarchy of communication is authoritarian, vertical, and one-way (parent-to-child) rather than cooperative, horizontal, collaborative, and reciprocal (adult-to-adult, human-to-human). The familiar irrational habits from our past usually preempt more reasonable adult-to-adult communication and block accurate associative reasoning processes, resulting in irrational thought, decision-making, and behavior. Parent-to-child-oriented irrational thought processes generally inhibit reasonable adult-to-adult communication, and this inhibition applies internally when we evaluate our own thoughts and emotions, as well as externally when we relate to others.

The strong tendency of humans to affiliate with groups and gravitate towards the familiar appears to impede adaptive change in humans and cultures, pulling the entire group backward toward more primitive, inaccurate, concrete ways of thinking, decision making, and behaving (Goetz & Shackelford, 2007, p. 13). The cultural inaccuracies we learn often contain an accumulation of past superstitious beliefs that dictate our behavior and soothe the angst of our limbic system (i.e. our emotions). Most cultural information results from implicit grammar and accumulates without the benefit of a scientific framework. Because we grow up under its influence, we generally fail to acquire the critical skills or awareness to assess the validity of our knowledge. Our innate affiliative behavior and inherited rigid thinking, coupled with inaccurate information, unfounded assumptions and lack of awareness, hinders our progress towards maximizing accurate rational thought and behavior, and more harmonious relationships.

Event-level Accuracy

Furthermore, it sets us up to "automatically" use outdated, ritualistic, familiar information and processes for solving a problem at a given moment, excluding new information and shifting us toward the past. Of course, old knowledge often has benefits—but if, by using it, we exclude new and possibly pertinent objective information, our familiar solutions may fail to adequately resolve problems we face in the present or future. We operate in the *now* as if it were identical to the *past*, creating a cognitive time warp. This rigid event-level distortion contributes to our irrational bias, magnified by our rigid, irrational, grammatical software.

Conversely, by selecting the most pertinent, accurate, and current information available and then processing it with flexible, accurate software, we shift our event-level orientation to the present, facilitating the most accurate, best-choice best-outcome decisions. More specifically, we then may use our frontal lobe executive function and working memory to make the most accurate rational choices by flexibly using all available pertinent objective information. In this way, we execute decisions in the present and develop plans for the future with a higher degree of precision and rational probability. This seems preferable to making irrational choices using irrelevant imprecise subjective information from the past, and then using the frontal lobes to retroactively justify and rationalize the decision. Accurate integration of time and space remains critical for problem solving and finding the best solution at any given time (Fuster, 2003, p. 62, 109).

Accurate Evaluations: Information and Processing

We would generally expect to achieve the best outcomes if the frontal lobes use all of the available timely and pertinent information to make decisions, not merely relying on what we learned in the past. This applies especially when we use higher-order executive cognition (Diamond, 2002, p. 494-95; Braver & Ruge, 2006, p. 307) to evaluate and integrate the emotional components of lower-order limbic cognition to differentiate between choices that feel good but may not serve our best interest, and choices that feel bad but may serve our best interest. Without accurate information about the situation at hand, we might decide on a course of action simply because it feels good or promises familiar rewards, or because it steers us clear of unknown or imagined threats. Furthermore, thinking accurately and rationally seems more likely to yield the best results when dealing with strong emotions, such as when interacting in bonded intimate relationships and affiliations, participating in territoriality and competition for resources, or when memories come with a particular emotion from the past that may represent an irrelevant concern of the present (Heilman, 1997, p. 135). Conversely, research shows that hormones (CRF) generated under stress tend to attenuate higher-order prefrontal cortex functioning, possibly shifting our thought processes towards lower level limbic and automatic implicit cognition. Therefore, we would probably benefit from applying the knowledge that rational higher-order cognitive functioning can preempt or minimize emotional stress and make a significant difference in outcomes. Cognitive therapies focus on rationally remodeling the grammar of belief systems, including percepts, concepts, and processing, to enhance objective awareness, teach stress management and coping skills, help to prevent or relieve depression and anxiety, and improve relationships (Beck, 1976, p. 328).

Benefits of Critical Thinking to Cognitive Accuracy

Critical thinking skills offer potential benefit for overcoming the disadvantages of normative uncritical thinking. Critical thinking encourages the adoption of more objective and scientific guiding principles useful for scrutiny of semantics, grammar, language, culture, assumptions, cause-effect conclusions, and conflict resolution. Critical thinking actually insists on questioning the totality of the argument, including the coherence of the argument process itself by emphasizing accurate, reliable and transparent evidence from a credible source, awareness of motivations and biases, clarity of observation and expression, and reliability of premises and assumptions, inferences, and conclusions. With critical thinking tools, we can evaluate the tactics used for irrational arguments by looking out for false beliefs, distractions, insults, catastrophe, perfect solutions, equivocation, appeal to popularity, authority, or emotions, distortions, false dilemmas, wishful-thinking, explaining by naming, hasty or glittering generalities, irrelevant topics, begging the question, etc.,(Kida, 2006; McInerny, 2005; Jones, 1998; Fisher, 2001; Aubyn, 1957).

Cultural Belief System and Constraints on Accurate Thinking

Current theory of normal human brain functioning indicates that we have the capacity to choose how and what we think, and we have a capacity to adapt to new situations by making better choices (Diamond, 2002, p. 494). By choosing objective information that is more accurate, and flexibly processing it, we increase our chances for determining the most accurate, reasonable, and timely solutions. In other words, we have the potential to bias our choices and outcomes flexibly, in a more accurate, objective way, incorporating semantic and information accuracy, accurate information processing, accurate time-appropriate and context appropriate information. Given this, why would we not choose an accurate rational bias with the highest degree of flexibility? Unfortunately, the subjective characteristics of our implicit grammar, along with personal, social, and cultural biases, impede us from understanding and using the very information that could enable such a choice (Beck, 1979, p. 13). Dichotomous grammar rigidly provides a potential fail-safe mechanism for perpetuating irrational implicitly learned cultural belief systems.

Because of these inherited belief systems, many humans tend to think they *must* be perfect, unblemished and without flaws, that they *must not* be flawed and fallible. A blemish or a mistake means they *are* no good, unworthy, or deserving of punishment. This absolute rating and labeling blocks recognition of personal fallibility, causes poor acceptance of others as fallible human beings, and promotes unscientific, culturally biased, arbitrary category classifications. This leads to rigid, judgmental, and dogmatic cultural belief systems, bigotry, stereotyping, and blind trust, especially blind trust in perceived authority. It also promotes vertical hierarchies with authoritarian, parent-to-child, one-way communication.

We often state our opinions as rigid, true-or-false, absolute statements about the universe, branding them as factual and either right or wrong. We use culturally determined, rigid, dichotomous, authoritarian, imperative, either-or terms, such as *should, must, have to*, and *need to*, implying that we have no other choices or that we *are obligated* to a certain choice. This cultural binding of *shoulds* tends to reinforce simplistic true-false perceptions, rigidity, and predetermination. Rigid, inaccurate, faulty beliefs and assumptions about the cause-effect relationship between thoughts, emotions, and behaviors inaccurately absolve us of responsibility for our own individual evaluations and decisions. Inaccurate definitions and rigid word use, combined with faulty cultural classifications, assumptions, generalities and absolutes, impede reasonable thought processes, hinder personal responsibility, and undermine accurate communication.

Grammatically constraining choices by defining them with rigid words and absolute concepts limits our cognitive options, especially in a *dynamic* world with many variables in a state of frequent change. We commit both a mathematical and a practical error when we arbitrarily restrict the use of multiple, possibly pertinent, variables and rigidly subscribe to culturally determined, dichotomous, *true or false* values when calculating choices and their outcomes (Langer, 2000; Langer & Piper, 1987; McInerny, 2005, p. 94-5). Even though our final choice may be binary, starting with the assumption of only prescriptive, non-contextual, binary choices imposes needlessly strict constraints. Rigidly held, inaccurate cultural belief systems decrease flexibility in our thinking, and limit our ability to develop harmonious relationships both as an individual and with others. These inaccuracies also perpetuate mind-brain dualism, a worldview that artificially separates the *mind* from the *brain* and fosters dichotomous, either-or thinking. Mind-brain dualism dates back to Plato and Descartes, and persists in many subsequent theories and beliefs (Morris & Dolan, 2004, p. 365). A recent study shows this fundamentally inaccurate worldview continues to thrive in the general population, ironically even among psychiatrists, psychologists, and mental health professionals (Miresco & Kirmayer, 2006).

Dichotomous bias, and the accompanying cognitive rigidity it fosters, exhibits many of the same features as dysexecutive functioning — inability to notice, integrate, or appropriately edit internal-external discrepancies in error detection and error correction (Braver & Ruge, 2006, p. 321). Rigid bias creates and perpetuates "reflexive" responses, similar to innate instinctive behaviors, with inflexible stimulus-response paradigms impervious to modification by context or experience (Mesulam, 2002 pp. 14-15, 22-26). Frontal lobe injury often interferes with divergent thinking, perhaps similar to the results one might expect with rigid cognition. Dysexecutive disorders fundamentally limit insight and bias cognition toward the use of primitive dichotomous grammar, inflexible behavior, limited learning from experience, perseveration, and concreteness (Milner & Petrides, 1984), appearing similar to primitive, poorly educated, fundamental cultures and belief systems. Evaluation of diagnostic categories might also benefit from a consideration of objective cognitive accuracy, given the apparent prevalence of dysexecutive features associated with many diagnostic categories. Disorders attributed to personality characteristics frequently present with dysexecutive components and such patients often have a history of developmental adversity and possibly other discrete frontal lobe dysfunction as well.

In addition, substance abuse, anxiety, adjustment disorders, marital discord, obesity and eating disorders may demonstrate unacknowledged frontal lobe dysfunction, possibly due to subjective classifications and biased, subjective, culture-bound heuristic evaluations and treatments (i.e., "we have always done it this way.") For example, consider a culturally biased definition of delusions as "false beliefs based on incorrect inference about external reality and firmly sustained in spite of the *opinion of others* or contrary evidence" (DSM). Evaluating one person's sanity by comparing it to the average opinions of others hardly sounds objective. How would one separate a delusion from prevalent subjective cultural beliefs or widespread superstitions using this definition? A dictionary definition seems a bit more clear, "a false belief or opinion resistant to reason or confrontation with actual fact" (Random House, 1997); however, this still leaves no easy way to distinguish delusions from opinions that differ from a culturally subjective, but factually invalid, consensus opinion. It seems that what many people call *facts*, i.e., beliefs and subjective knowledge, actually serve as subjective grammatical reference points that vary from culture to culture without regard for objective accuracy. How do we distinguish the "cultural facts", i.e., the imagined certainties of static subjective beliefs, from observed, uncertain, probabilistic, dynamic, scientific facts?

Reference Points for Cognitive Accuracy and Rational Bias

Life unfolds as a series of choices and outcomes. We would like to predict the outcome of a particular choice with some degree of certainty, but doing so depends on understanding the many variables in our ever-changing, *dynamic* world. It seems important in this complicated dynamic environment, to accurately assess probable outcomes using flexible, multivariate thought processing and to adequately evaluate our many choices and desired consequences, increasing our contextual response to a given situation. This enhances our overall adaptability. "Choices can be rational or they can be the outcome of irrational processes" (Benjafield, 2007, p.4). A deliberate bias towards rationality tends to enhance our overall accuracy. A rational dynamic bias favors more effective decision-making and increases the probability of more reasonable outcomes. On the other hand, an irrational static bias tends to decrease overall accuracy, leading to irrational decision-making with fewer reasonable choices and fewer objective outcomes. The standards for measuring accurate and rational cognitive bias arise in part from the following assumptions:

- *Acceptance of human imperfection enhances information and process accuracy.* We can accurately characterize humans as flawed and fallible. Accepting our own flaws and fallibilities encourages us to accept others as human beings, albeit with flaws. This acceptance promotes horizontal, human-human, adult-adult collateral communication. It also reduces inaccurate, absolute, dichotomous, or culturally biased classifications, ratings, and labeling, because we do not believe that any person is *all* bad or *all* good. It seems much more reasonable to rate the behavior rather than the person. Rational human acceptance minimizes inaccurate judgmental categories and cultural bigotry while promoting realistic scientific belief systems and healthy skepticism.
- *Flexibility enhances information and process accuracy.* Flexibility generally works better than rigidity for the most accurate planning, problem solving, and compromising in a dynamic world. Rigid, dichotomous, culturally determined terms like *should, must, have to, got to*, and *need to*, restrict options and diversity, while multivariate, preferential terms, such as *I would prefer, I would rather*, and *to me, this*

seems best, multiply the diversity of possible choices and acceptable rational outcomes. Opinions and preferences replace absolute declarations of right and wrong. Generalities stated as assumptions or deductions represent a higher level of accuracy than generalities misrepresented as true or false *facts* when they actually represent subjective *beliefs*. Such generalizations amplify inaccuracies. Moreover, we improve the precision of our thought processes, decisions, and communications by using the most accurate word definitions, making specific rather than vague statements, using multivariate spectra or gradients for evaluations and revaluations, and avoiding faulty cause-effect conclusions (Browne & Keeley, 2007, p. 147).

- *Awareness of the relationship between thoughts and emotions enhances information, process, and event-level accuracy.* Normally, cognition has a significant causal relationship with our feelings, and realizing this enhances accurate assessment of individual responsibility for thoughts, feelings, and behaviors. Our thoughts, or the rules behind our thoughts, cause or significantly influence our feelings whether we recognize the connection or not. Awareness of this relationship enables us to choose the healthiest and most rational thoughts in order to maximize our emotional and behavioral balance at a given time. Although we might initially react to the situation itself, we largely generate and sustain our emotional reactions to events by what we think or "believe" about them (Ochsner, 2006, p. 245-50). We tend to sustain the emotion long afterwards through the action of implicit internal rules and appraisal habits that affect us almost continuously, generally without our awareness or deliberate direction. As Epictetus wrote in the *Enchiridion* in the first century, "People are disturbed not by things, but by the views which they take of them" (Ellis & Harper, 1997, p. 39).

Cognitive Awareness

Becoming aware of our internal narratives, rules and beliefs about the world and events gives us some understanding of the effect we have on our own individual state and, with that understanding, the ability to objectively moderate our reactions. When something unexpected happens, we often tell ourselves inaccurate, irrational, and overly negative things about the situation, needlessly upsetting and stressing ourselves about it. If instead we insightfully choose to describe the situation as accurately as possible, we can respond with the most appropriate behavior and most reasonable emotion. "Why *must* you upset your self?" (Albert Ellis, attributed). We do not *have to* upset ourselves needlessly or act irresponsibly on irrational thoughts. We improve our individual accountability when we take responsibility for our thoughts and how those thoughts affect our feelings and behaviors. We each have responsibility for managing our own cognitive accuracy, including our individual responsibility for our thoughts, emotions, and behaviors. This enhances reasonable behavior and harmony between humans (Gemba, 2002, p. 146).

Emotional disturbance, in sum, usually stems from your Irrational Beliefs. You can uncover the basic unrealistic ideas with which you disturb yourself; see clearly how misleading these ideas are; and, on the basis of better information and clearer thinking, *change* the Beliefs behind your disturbance (Ellis & Harper, 1997, p. 69).

An orientation towards cognitive accuracy encourages active, objective thinking in the present with proactive, forward-looking, active, event-level evaluations; adaptability; continuous quality improvement; and positive reinforcement and recognition of the importance of rational thought processes. The complexity of life magnifies the importance of thinking accurately. Acceptance of human imperfection, bias towards flexibility, and individual responsibility enhance the accuracy and quality of our cognition and increase the probability of achieving the most reasonable (i.e., *best*) outcomes. This improved quality of accurate thought promotes more harmonious interactions within and between individuals. It also promotes awareness of the benefits of thinking and acting rationally, completing the circle:

It is not the strongest of the species that survive, nor the most intelligent, but the one most responsive to change (Commonly attributed to Charles Darwin, biologist).

Correcting Irrational Biases

However, how can we learn these new skills if, throughout our development, we see few examples of critical thinking and adults fail to teach us to think more logically? How do we improve our cognitive accuracy when we receive little or no foundation for rational or logical information templates in our memory storage areas to build on? Finally, how can we use our acquired, almost "fail safe," irrational, rigid, dichotomous thought processes to learn how to think rationally and make choices that are more rational?

Fortunately, we have the capacity to replace inaccurate, irrational thinking habits with newer, more accurate, rational associative thought processes. It takes effort and practice to learn and use new concepts and retool our brain library and dictionary (Lieberman, 2006, p. 201) with new information and software. The greater the effort (i.e., the more often and harder we practice), the sooner we can replace our habitual, inaccurate, irrational thinking with the new objective skill of accurate rational thinking. This objective shift in the way we think uses higher-level, multivariate, associative reasoning instead of rigid, black-and-white, either-or cognition to evaluate choices and outcomes. Rational associative reasoning tends to maximize accurate executive cognition and decision-making. This increased accuracy contributes directly to thinking more rationally with more reasonable outcomes, because our higher-level executive functioning has the utility of the "last word."

Implications of Irrational Bias

> "Our present problems cannot be solved at the level of thinking at which they were created" (Albert Einstein, attributed).

When applied to decision-making, rigid, irrational biases tend to employ a floating reference point; in contrast, biases that are more flexible and rational tend to operate from a stable reference point. How can we explain this seeming paradox? Understanding it depends on the perspective of the observer. An observer using the same irrational biases, grammar, terms, and definitions used to construct the irrational reference point will find this conundrum baffling. As previously stated, humans rigidly cling to their cultural belief systems, no matter how subjectively antiquated or irrelevant. These belief systems take hold during early learning, along with the related dominance characteristics of parent-child interactions and obedience to authority (Milgram, 2004, pp.114-5, 136-47). We learn, whether directly or indirectly, that *good* is rewarded and *bad* is punished, that *right* is rewarded and *wrong* is punished, and that parents know what is *right* and what is *wrong*. The problem arises when we apply these static dichotomous grammatical constructs to the complexity of the dynamic variables of the world we live in. Humans tend to perceive the rules for decision making in hierarchal, dichotomous and absolutistic terms: is, is not, black, white, should, must, have to, ought to, etc. *You must do that or you will be punished.* This creates a problem when the must of today becomes the must not of tomorrow because some variable or context has changed.

Reference Point Drift

This dichotomous grammar then leads to retrospective judging, faultfinding, blaming, and punishing. *You should not have done that*! This rationale might work if we had a means of predicting the future in our dynamic environment; given that we do not, we look for another mechanism to explain the unwelcome outcome. To resolve the apparent discrepancies with reality, we simply move our reference point: We redefine success or modify our recollection of errors to cast the current outcome in a better light (Schlacter, 2001, p. 151). By moving the reference point, we can always be right, always be above average, and never make a mistake. We simply move the reference point to under-value others or over-value ourselves, and this creates the irrational but comforting experience of "self-esteem." We perhaps learn this skill in childhood when governed by the multiple, subjective, and inconsistent reference points of parents, teachers, neighbors, friends, celebrities, etc.

We also learn to easily construct subjectively biased, over-generalized groups and classes and use them to denigrate and subjugate others at will. We can easily over-generalize or under-define representative classifications and prove most any point using imprecise labels. We can round up or down at will. We have our choice of measuring instruments and usually tend to choose the yardstick that casts us in the most favorable light, with little thought about accuracy. Since each individual chooses their own rules from their own frame of reference, we can use one rule to justify and a different rule to vilify any one experience to suit our own beliefs and goals (Kida, 2006, p. 78), thereby potentially artificially elevating self-esteem. This works well for confirmation bias (Kida, 2006, p. 160).

Social cognitive neuroscience has demonstrated that when humans see others as humans, they tend to moderate their prejudices (Mitchell et al., 2006) and that positive experiences with others can erase learned cultural stereotypes (Phelps & LaBar, 2006, p. 440). Milligan found that when volunteer experimenters, under commands of authority, viewed experimental subjects as "unworthy persons" rather than humans, their willingness to punish them increased (Milligan, 2004, p. 161). In another study, volunteer experimenters punished subjects less when the experimenter was in closer proximity to subjects (Milligan, 2004, pp. 37). This suggests that subjective human labeling contributes to prejudice and punishment by making others less than human. When we remove the subjective labels and see others objectively as humans, like us, we can acknowledge that we belong to the same class, *Homo sapiens*. We may not approve of certain behaviors in others but we have the option of labeling the behavior rather than the person.

Humans often blame others for their own thoughts, emotions, and behaviors. Since we fear being wrong, we simply decide that someone else *made* us think, feel, or behave a certain way. We shift the blame to a scapegoat. They are at fault. They are to blame. Simply by adjusting the reference point, we easily pass along the blame and judge ourselves not guilty. This approach may succeed in bolstering the "self-esteem" of the blamer, but usually at the expense of anger or guilt of the blamed, due to their resentment or internally demoted *self-esteem*. This can have disastrous effects on long-term relationships.

Irrationality of Self-Esteem versus Rational Acceptance of Human Imperfection

Unfortunately, self-esteem represents an authoritative artifact of irrational cultural constructs that depend on dichotomous judgments of right and wrong, perfect and imperfect, and other culturally defined, rigid, subordinate subjective values (Milligan, 2004, p. 147). A stable reference point based on cognitive accuracy uses a consistent spectrum to gauge behavioral achievement, performance, and error, in effect giving us standards that provide a more rational reality for testing outcomes, while avoiding statements about individual worth. With a rational bias, we accept that a certain rate of failure is inevitable; this minimizes the motivation to "fix" our mistakes by justifying, rationalizing, or blaming others. This, in turn, helps us resist the temptation to float the mean or tweak the reference point to make our decisions appear more above average or always correct. We can freely admit our error and concentrate on improvement.

Acceptance of human imperfection tends to be a more objective and workable construct than self-esteem (Ellis, 2005) because it starts with reality, the rational assertion that human beings have flaws and make fallible decisions, and therefore err on occasion. If we embrace human fallibility and the uncertainty of our choices, we do not suffer an intolerable shock when we fail, and we do not find it necessary to change history or redefine our success and achievements to feel happy. However, even if we rationally accept the premise that we will rarely predict the future with 100% accuracy, we still face a potential challenge from the parts of our brain that store and apply our cultural beliefs and irrational biases. As the gatekeeper of the limbic system, in charge of safety, the amygdala appears to continue to preferentially reward the perceived certainty of familiar choices to such an extent that we still routinely choose against our own best interests. We even chisel our principles and beliefs in stone and refuse to admit to any degree of error. This mollifies the amygdala, when operating on irrational beliefs with black and white rigidity in a mostly gray world. Then, instead of rationally evaluating and predicting the future, we habitually use our exceptional frontal lobes to *rationalize the past*. We make up after-the-fact explanations and excuses to *justify* the errors of our behaviors, refute reality, or recast history according to our own wishes while magically preserving our self-esteem. Fortunately, the amygdala responds to cognitive control (Phelps, 2004, p. 1013). Cognitive and rational therapies temper the irrational anxiety of the amygdala by providing education in objectivity and competent rational strategies.

Our human tendency to rationalize and justify may come from our evolutionary roots. As our primitive semantics, grammar, and language arose from symbolic grunts, gestures, and facial expressions to form the earliest beginnings of more formal communication, we began to use words to label objects and actions (rock...not rock, bad...not bad). From this, dichotomous grammar evolved. This may have worked well in primitive cultures to improve group adaptation to the environment, using dominance, punishment, reward for accomplishments, and faulty interpretation of coincidences extrapolated to causation. The *veridical* nature of black and white, dichotomous grammar increases the probability of error in the mostly gray and ever-changing *dynamic* world we live in. Ironically, these increased errors of dichotomous cognition also increase deviation from the mean, along with increased variability in dichotomous rule adherence. These errors and deviations cause internal-external evaluation discordance. The easiest irrational corrective solution involves generating explanations for these dissonant discrepancies in the form of irrational beliefs, heuristics, and their accompanying biases.

In early primitive cultures that lacked scientific knowledge, language arose out of reasonable attempts to label objects and understand relationships between humans and the environment initially using nouns (objects) and verbs (action). Since nouns, verbs, and language predated writing, we might reasonably assume they sustained an early narrative historical record and that progress in objective knowledge made it increasingly more difficult to explain away uneducated, unscientific beliefs, faulty labeling, and transfer of fault and blame to others. However, heuristics may have evolved as a form of mental shortcuts that were then extrapolated to justifications for irrational beliefs, thoughts, and behavior. Heuristics represent easily stated and applied "rules of thumb" that develop as a part of most cultural belief systems. While they generally purport to convey compact bits of wisdom, they more often embody cognitive and verbal tools for evading reality. They enable humans to correct errors by explaining them away, by generalizing, justifying, dignifying, signifying, rationalizing, blaming, bullying, revising, labeling, etc. We can then resort to the prefabrication of even more *preposterous* irrational rules somewhat akin to confabulation.

With irrational, dichotomous grammar, humans can create a scorecard, noting their own successes and ignoring their own errors, while tracking the errors of others and ignoring their successes. This allows us to maintain a perception of "self-esteem," and remain above the mean, above average and perfect. Heuristics and irrational beliefs make it easier to evaluate internal-external errors and error corrections from our own frame of reference. Unfortunately, dichotomous grammar increases errors and decreases error detection, whether we can explain them away or not. Objective multivariate grammar generally makes more efficient use of the limited storage space afforded to humans (Nowak, 2006, pp. 249-86) than irrational grammar can. Rational multivariate grammar seems to offer increased accuracy without the convoluted perturbations usually required to correct for errors after the fact under subjective dichotomous grammar. It would seem plausible then that rational, multivariate grammar and acceptance of human imperfection offers a more efficient and a more reasonable construct than irrational grammar and self-esteem.

DISCUSSION, CONCLUSIONS, AND CONSIDERATIONS

We have the resources to counter these mostly culturally determined irrational and inaccurate grammatical habits by diligent pursuit of accurate cognition, even in the face of doubt and uncertainty. We can retrain the brain to value rational, probabilistic outcomes, accept human imperfection in place of self-esteem, and use accurate information and processing to make decisions about current situations. Each of these practices can eventually enable us to use the processing capabilities of our new brain to direct the older, survival-oriented brain structures (MacLean, 1990; Milgram, 2004) toward more rational pursuits. We can then focus on achieving objective improvement and obtaining better outcomes for the future (Panksepp, 1998, p. 260). This suggests that establishing reference points for cognitive accuracy may provide a naturalistic rational framework for future research that will demonstrate an ecologically valid unifying theory for brain and behavior in *Homo sapiens*. Cognitive accuracy appears to represent an evolution-friendly *multivariate* grammar that might impart greater homeostatic benefits. Objective multivariate grammar meets requirements for simplicity (Occam's razor), and seems to offer a more efficient homeostatic mechanism, possibly minimizing the liabilities incurred by the inherent inaccuracies of subjective dichotomous grammar. Thus, cognitive accuracy may provide useful fundamental principles for establishing a scientific reference point for the evaluation of rational behaviors, choices, and consequences. A potential scientific method based on cognitive accuracy includes three fundamental principles:

- information accuracy: seeking and using objective information based on empirical observation; premise, deduction, conclusions, and testing
- thought process accuracy: making evaluations and decisions flexibly with critical thinking, multivariate terminology, and awareness of individual responsibility
- event-level accuracy: connecting and verifying both information and decisions in a time- and context-dependent manner to increase the relative probability of more accurate predictions of rational outcomes (Bailey, 2007, in press; Bailey, 2006)

Each component helps to ensure reliable cognitive functioning. Event-level accuracy is especially important for maximizing context specificity by accurately integrating cognition (information and information processing), and behavior at a given moment and place. This identifies, acknowledges, and "preserves an independent reality" (Minkowski, 1952, p. 76).

The next shift in human primate cognitive evolution could potentially consist of the implementation and integration of cognitive accuracy into all major areas of scientific study, thereby fostering individual and cultural acceptance and direct application. If so, then thinking more flexibly, accepting uncertainty, and living rationally in the present may become the norm instead of the exception. Increasing our degree of accurate, rational thinking would appear to enhance our prospects for living rationally in the present and planning for a more reasonable and likely future. As accurate, rational thoughts and behaviors increase, irrational thoughts and behaviors tend to decrease. Thinking reasonably and rationally relies on the accuracy of higher-order executive functioning, with the potential to transport humankind to a more human and humane future (Hendelman, 2006, p. 238).

Accurate, rational thinking increases flexibility and tends to maximize appropriate choices, enhancing achievement of preferred outcomes while minimizing undesirable irrational outcomes, without the liabilities of irrational cognition (Fine, 2006, p. 208). Fortunately, we have the ability to replace inaccurate, irrational, parental, absolute thinking by learning and practicing new habits of accurate, flexible, rational, and reasonable logical critical thinking, thus improving overall adaptability. We have these rational tools available, but they often go unrecognized, overlooked, or even belittled (Jones, 1998, pp. 6-7).

Ideally, we will achieve this accurate, rational, cognitive evolutionary step before we trigger an irrationally induced catastrophe with our current habits of rigidity, disharmony, self-loathing, and aggression (Beck, 1999).

Science has understood the negative effects of subjective cultural beliefs on the evaluation of objective knowledge for some time. How we measure something in large part defines what we measure. "What you get from a measurement depends on what you choose to measure" (Lindley, 2007, p. 155). We tend to calibrate our personal yardsticks to the culture in which we find ourselves. Such a yardstick measures only what that culture values. Each culture's yardstick shows normal, despite large differences between the beliefs and behaviors of various cultures. This potentially leaves us prey to the accumulated inaccuracies of our forebears.

Cognitive accuracy represents a reference point biased toward objective cognitive accuracy in order to measure human thought and behavior across cultures. This reference point does not change as you go from one culture to the next, theoretically transcending the inaccuracies of cultural belief systems. Cultures may change, but the yardstick remains the same, unless and until scientific advances using cognitive accuracy indicate beneficial objective adjustments. For accurate evaluations, we do best to calibrate our cognitive yardstick with the most accurate, timeliest information, applied consistently and rationally in the present.

This article has reviewed the intimate connection between semantics, grammar, cognition, and the functioning of various brain structures, and posited an equally intimate association between semantics, grammar, cognition, emotion, and behavior. These associations suggest an explanation for the persistence of subjective cultural belief systems and their inherent deficiencies in objectivity, accuracy, and flexibility. Can we accept the premise that objective, timely, accurate processing and accurate information usage might efficiently yield more satisfactory and reasonable results than rigid automatic responses based on inherited imprecise beliefs? If so, it would seem beneficial to society for science to orient new research to incorporate the psychological value of objective relative belief systems, using scientific insights from cognitive neuroscience, anthropology to zoology, ethology to neuropsychology, cognitive behavioral science to neuroeconomics, affective neuroscience, evolutionary dynamics etc., along with the tools of cognitive accuracy.

Cumulative prospect theories and neuroeconomics (Kahneman & Tversky, 1992; Glimcher et al., 2005) suggest that humans weigh the relative probabilities of achieving a particular outcome along an economic gradient of values that influence cognition, emotions, and behavior. Our tendency to seek reward and familiarity and generally avoid risk, uncertainty, and pain, combined with the inertia of our acquired cultural belief systems, biases our value analysis and skews choice-outcome decisions disproportionately in favor of the familiar or the presumed socially desirable outcome (Tom et al., 2007). Moreover, cultural beliefs are resistant to transformation once established (Panksepp, 1998, p. 245; Lane et al., 2000, p. 409), suggesting that whatever adult behavior (including cognition) we wish to promote in a society, we will do best to teach during childhood development and human maturation (Tse et al., 2007). For many generations, parents have raised children to hold the same rigid, unexamined beliefs they hold themselves. If we instead model humane treatment of others as humans, flexibly apply objective critical thinking, and demonstrate reasonable behavior for our children, we have some hope they will instead value cognitive accuracy as adults. Luria, many years ago, found evidence that early education can teach children the ability to use logical thought (Luria, 1981, p. 209). This suggests that exposure to and education in higher-level logical reasoning, throughout development, could enhance rational cognition and behavior.

Once a sufficient number of adults acquire the skills and habit of cognitive accuracy, their interactions with children would likely pass on fewer faulty beliefs and thought processes. Importantly, at an early age, children raised by these adults would have the opportunity to learn to develop and extend their cognitive accuracy. As adults, they would have more skills for applying cognitive accuracy to address and resolve problems with competent critical thinking, emotional balance, and reasonable behaviors. This would enable each successive generation to build on the successes of their parents, rationally living in the present (Korzybski, 1958, p. 231).

Science, by virtue of its central methodology (Browne & Keeley, 2007, p. 119), speaks for the value of accurate, rational, logical objective reasoning and critical thinking, and for advocating the teaching of objective rational cognition, which can potentially guide society towards increasingly humane thought and behavior. Future research on human behavior and neuroscience can contribute to this goal by taking into account the interdependence of biology and society. If we have the choice to conduct research that grounds our cultural belief systems in neuroscience, and it seems we do, while producing more reasonable and responsible adults who use cognitive accuracy and rational evaluations to make decisions, can we afford not to do so?

In the final analysis, we have to depend on our rich resources of rationality to recognize and modify our irrationality…We can recognize that our own interests are best served by applying reason. In this way, we can help to provide a better life for ourselves, others, and the future children of the world, (Beck, 1999, p. 287).

AUTHOR NOTES

Charles E. Bailey, M.D. is a General Psychiatrist and Clinical Research Psychopharmacologist in Orlando, Florida. cbailey1@cfl.rr.com

ACKNOWLEDGMENTS

A special thanks to Nora Miller for her tireless editing efforts; posthumously to Albert Ellis for his editorial comments and encouragement; to Elkhonon Goldberg for his encouragement and sage advice; to M.-Marsel Mesulam for his very thoughtful suggestions; to Khrista Boylan for her valuable input; and to Scott Rauch and Carl Senior for taking time out of their busy schedules to make much appreciated and insightful editorial comments.

REFERENCES

Adolphs, R. (2006). What is special about social cognition? In J. T. Cacioppo, P. S. Visser, & C. L. Pickett (Eds.), *Social neuroscience, people thinking about people* (pp. 269-285). Cambridge, MA: MIT Press.

Alexander, G. E., Delong, M. R., & Strick, P. L. (1986). Parallel organization of functionally segregated circuits linking basal ganglia and cortex. *Annual Review of Neuroscience*, 9, 357-381.

American Psychiatric Association: *Diagnostic and Statistical Manual of Mental Disorders*, Fourth Edition. Washington, DC: American Psychiatric Association, 1994.

Anderson, V., Levin, H. S. & Jacobs, R. (2002). Executive functions after frontal lobe injury: A developmental perspective. In D. T. Stuss & R. T. Knight (Eds.), *Principles of frontal lobe function* (pp. 504-527). New York, NY: Oxford University Press.

Aubyn, G. St. (1957). *The art of argument*. USA: Emerson Books.

Avrutin, S. (2006). Weak Syntax. In Y. Grodzinsky & K. Amunts (Eds.), *Broca's region* (pp. 49-61). New York, NY: Oxford University Press.

Baddeley, A. (2002). Fractionating the central executive. In D. T. Stuss & R. T. Knight (Eds.), *Principles of frontal lobe function* (pp. 246-259). New York, NY: Oxford University Press.

Bailey, C. E. (2006). A general theory of psychological relativity and cognitive evolution. *ETC: A Review of General Semantics*, 63, 278-289.

Bailey, C. E. (2007) in press. Cognitive accuracy and intelligent executive functioning in the brain and in business. In C. Senior & M. J. R. Butler (Eds.), *The social cognitive neuroscience of business*. New York, NY: Annals of the New York Academy of Sciences.

Barbas, H. (2006). Organization of the principle pathways of prefrontal lateral, medial, and orbitofrontal cortices in primates and implications for their collaborative interaction in executive functions. In J. Risberg & J. Grafman (Eds.), *The frontal lobes, development, function and pathology* (pp. 21-68). New York, NY: Cambridge University Press.

Bartel, A. & Zeki S. (2004). The neural correlates of maternal and romantic love. *Neuroimage*, 21, 1155-1166

Baum, E. B. (2004). *What is thought?* Cambridge, MA: MIT Press.

Beck, A. T. (1999). *Prisoners of hate: The cognitive basis of anger, hostility, and violence.* New York, NY: HarperCollins.

Beck, A. T. (1976). *Cognitive therapy and the emotional disorders.* New York, NY: International University Press.

Benjafield, J. G. (2007). *Cognition* (3rd Ed.). Canada: Oxford University Press.

Bloom, P. (2002). *How children learn the meaning of words.* Cambridge, MA: MIT Press.

Braver, T. S. & Ruge, H. (2006). Functional neuroimaging of executive functions. In R. Cabeza & A. Kingstone (Eds.), *Handbook of functional neuroimaging of cognition* (pp. 307-348). Cambridge, MA: MIT Press.

Browne, M. N. & Keeley, S. M. (2007). *Asking the right questions: A guide to critical thinking* (8th ED.). CITY? NJ: Pearson Prentice Hall

Boyd, R. & Richerson, P. T. (2005). *The origin and evolution of cultures* (p. 206). New York, NY: Oxford University Press.

Broca, P. (1877). Rapport sur un memorie M. Armund de Fleury intitule: De l'inegalite dynamique des deux hemispheres cerebraux. *Bull Academy of Medicine*, 6, 508-539.

Cacioppo, J. T. & Berntson, G. G. (2004). Social neuroscience. In M. S. Gazzaniga (Ed.), *The cognitive neurosciences* III (pp. 977-985). Cambridge, MA: MIT Press.

Caldwell, H. K. & Young III, W. S. (2006). Oxytocin and vasopressin: Genetics and behavioral implications. In R. Lim (Ed.) *Handbook of neurochemistry and molecular neurobiology: Neuroactive proteins and peptides*, 3rd Ed. (pp. 573-607). New York, NY: Springer

Capaldi, N. (1987). *The art of deception: An introduction to critical thinking* (p.17). New York, NY: Prometheus Books.

Cho, M. M., Devries, A. C., Williams, J. R. & Carter, C. S. (1999). The effects of oxytocin and vasopressin on partner preferences in male and female prairie voles. *Behavioral Neuroscience*, 113, 1071-1079.

Chow, T. W. & Cummings, J. L. (2007). Frontal-Subcortical circuits. In B. L. Miller & J. L. Cummings (Eds.), *The human frontal lobes*, 2nd Ed. (pp. 25-43). New York, NY: Guilford Press.

Cummings, J. L., & Miller, B. L. (2007). Conceptual and clinical aspects of the frontal lobes. In B. L. Miller & J. L. Cummings (Eds.), *The human frontal lobes*, 2nd Ed. (pp. 12-21). New York, NY: Guilford Press.

Curtis, C. E. & D'Esposito, M. (2006). Functional neuroimaging of working memory. In R. Cabeza & A. Kingstone (Eds.), *Handbook of functional neuroimaging of cognition* (pp. 269-306). Cambridge, MA: MIT Press.

Davachi, L., Romanski, L. M., Chafee, M. V. & Goldman-Rakic, P. S. (2004). Domain specificity in cognitive systems. In M. S. Gazzaniga (Ed.), *The cognitive neurosciences* III (pp. 665-678). Cambridge, MA: MIT Press.

Damasio, A. R. (2000). A second chance for emotion. In R. D. Lane & L. Nadel (Eds.), *Cognitive neuroscience of emotion* (pp. 12-23). New York, NY: Oxford University Press.

Deardorff, A. V. (2006). *Terms of trade: Glossary of international economics*. Hackensack, NJ: World Scientific Publishing Co.

Decety, J. (2007). A social cognitive neuroscience model of human empathy. In E. Harmon-Jones & P. Winkielman (Eds.), *Social neuroscience: Integrating biological explanations of social behavior* (pp. 246-270). New York, NY: Guilford Press.

Delgado, M. R. (2007). Reward-Related responses in the human striatum. In B. W. Balleine, K. Doya, J. O'Doherty & M. Sakagami (Eds.). Reward and decision making in corticobasal ganglia networks. *Annals of the New York Academy of Sciences*, 1104, 70-88.

Dennis, M. (2006). Prefrontal cortex: Typical and atypical development. In J. Risberg & J. Grafman (Eds.), *The frontal lobes, development, function and pathology* (pp. 128-162). New York, NY: Cambridge University Press.

D'Esposito, M. & Postle, B. R. (2002). The organization of working memory function in lateral prefrontal cortex: Evidence from event-related functional MRI. In D. T. Stuss & R. T. Knight (Eds.), *Principles of frontal lobe function* (pp. 168-187). New York, NY: Oxford University Press.

Diamond, A. (2002). Normal development of prefrontal cortex from birth to young adulthood: Cognitive functions, anatomy, and biochemistry. In D. T. Stuss & R. T. Knight (Eds.), *Principles of frontal lobe function* (pp. 466-503). New York, NY: Oxford University Press.

Duvernoy, H. M. (2005). *The human hippocampus* (3rd Ed.). New York, NY: Springer-Verlag Berlin Heidelberg.

Edelman, G. M. (1992). *Bright air, brilliant fire: On the matter of the mind.* New York, NY: Basic Books.

Ellis, A. (2005). *The myth of self-esteem: How rational emotive behavior therapy can change your life forever.* Amherst, NY: Prometheus Books.

Ellis, A. & Harper R.A. (1997). *A guide to rational living* (3rd Ed.). N. Hollywood, CA: Melvin Powers Wilshire Book Company (Original work published 1976).

Ferris, C. F. (2006). Neuroplasticity and aggression: An interaction between vasopressin and serotonin. In R. J. Nelson (Ed.), *Biology of Aggression* (pp. 163-175). New York, NY: Oxford University Press.

Ferstl, E. C. (2007). The functional neuroanatomy of text comprehension. In F. Schmalhofer & C. A. Perfetti (Eds.), *Higher level language process in the brain: Inference and comprehension processes* (pp. 53-102). Mahwah, NJ: Lawrence Erlbaum Associates Inc.

Fine, C. (2006). *A mind of its own: How your brain distorts and deceives* (p. 208). New York, NY: Norton.

Fisher, A. (2001). *Critical thinking: An introduction.* Cambridge, UK: Cambridge University Press.

Fisher, H. E., Aron, A., Mashek, D., Li, H. & Brown, L. L. (2002). Defining the brain systems of lust, romantic attraction, and attachment. *Archives of Sexual Behavior*, 31, 413-419.

Frith, C., Rees, G., Macaluso, E. & Blakemore, S. (2004). Higher cognitive functions: Mechanisms of attention. In R. S. J. Frackowiak, K. J. Friston, C. D. Frith, R. J. Dolan, C. J. Price, S. Zeki, J. Ashburner & W. Penny (Eds.), *Human Brain Function* (pp. 245 -268) (2nd Ed). San Diego, CA: Elsevier Science USA.

Fuster, J. M. (2003). *Cortex and mind, unifying cognition.* New York, NY: Oxford University Press.

Gazzaley, A. & D'Esposito, M. (2007). Unifying prefrontal cortex function: Executive control, neural networks, and top-down modulation. In B. L. Miller & J. L. Cummings (Eds.), *The human frontal lobes*, 2nd Ed. (pp. 187-206). New York, NY: Guilford Press.

Gemba, H. (2002). Motor programming for hand and vocalizing movements. In D. T. Stuss & R. T. Knight (Eds.), *Principles of frontal lobe function* (pp. 127-148). New York, NY: Oxford University Press.

Giesbrecht, B., Kingstone, A., Handy, T. C., Hopfinger, J. P. & Mangun, G. R. (2006). Functional neuroimaging of attention. In R. Cabeza & A. Kingstone (Eds.), *Handbook of functional neuroimaging of cognition* (pp. 85-111). Cambridge, MA: MIT Press.

Glimcher, P. W., Dorris, M. C. & Bayer, H. M. (2006). Physiological utility theory and the neuroeconomics of choice. *Games and Economic Behavior*, 52, 213-256.

Goetz, G. T. & Shackelford, T. K. (2007). Introduction to evolutionary theory and its modern application to human behavior and cognition. In S. M. Platek, J. P. Keenan & T. K. Shackelford (Eds.), *Evolutionary cognitive neuroscience* (pp. 5-19). Cambridge, MA: MIT Press.

Goldberg, E. (2002). *Executive brain: Frontal lobes and the civilized mind.* New York, NY: Oxford University Press.

Goldman-Rakic, P. S. (1995). Architecture of the prefrontal cortex and central executive. *Annals of the New York Academies of Science*, 769, 71-83.

Grafman, J. (2002). The structured event complex and the human prefrontal cortex. In D. T. Stuss & R. T. Knight (Eds.), *Principles of frontal lobe function* (pp. 292-310). New York, NY: Oxford University Press.

Graybiel, A. M. & Saka, E. (2004). The basal ganglia and the control of action. In M. S. Gazzaniga (Ed.), *The cognitive neurosciences* III (pp. 495-510). Cambridge, MA: MIT Press.

Hauser, M. D., Chomsky, N. & Fitch, W. T. (2002). The faculty of language: What is it, who has it, and how did it evolve? *Science*, 298, 1569-1579.

Hebb, D. O. (1949). *The organization of behavior.* New York, NY: Wiley.

Heilman, K. M. (1997). The neurobiology of emotional experience. In S. Salloway, P. Malloy & J. L. Cummings (Eds.), *The neuropsychiatry of limbic and subcortical disorders* (pp. 133-142). Washington, DC: American Psychiatric Press.

Heimer, L., Alheid, G. F., de Olmos, J. S., Groenewegen, H. J., Haber, S. N., Harlan, R. E. & Zahm, D. S. (1997). The Accumbens: Beyond the core-shell dichotomy. In S. Salloway, P. Malloy & J. L. Cummings (Eds.), *The neuropsychiatry of limbic and subcortical disorders* (pp. 43-70). Washington, DC: American Psychiatric Press.

Hendelman, W. J. (2006). *Atlas of functional neuroanatomy* (2nd Ed.). Boca Raton, FL: CRC Press.

Hooker, C. I. & Knight, R. T. (2006). The role of lateral orbitofrontal cortex in the inhibitory control of emotion. In D. H. Zald & S. L. Rauch (Eds.), *The orbital frontal cortex* (pp. 307-324). New York, NY: Oxford University Press.

Insel, T. R. (1997). A neurobiological basis of social attachment. *American Journal of Psychiatry,* 154, 726-735.

Jones, M. D. (1998). *The thinker's toolkit: 14 powerful techniques for problem solving* (p. 6-7, 14-23, 33, 46). New York, NY: Three Rivers Press.

Kahneman, D., Slovic, P. & Tversky, A. (1982). *Judgment under Uncertainty: Heuristics and Biases.* Cambridge, MA: Cambridge University Press.

Kaufer, D. L. (2007). The dorsolateral and cingulate cortex. In B. L. Miller & J. L. Cummings (Eds.), *The human frontal lobes*, 2nd Ed. (pp. 44-58). New York, NY: Guilford Press.

Keller, H & Chasiotis, A. (2006). Evolutionary perspectives on social engagement. In P. T. Marshall & N. A. Fox (Eds.), *The development of social engagement: Neurobiological perspectives* (pp. 275-303). New York, NY: Oxford University Press.

Kida, T. E. (2006). *Don't believe everything you think: The 6 basic mistakes we make in thinking.* Amherst, NY: Prometheus Books.

Korzybski, A. (1958). *Science and sanity: An introduction to non-aristotelian systems and general semantics* (4th Ed.). Lakeville, CT: International Non-Aristotelian Library Publishing (Original work published 1933).

Langer, E. J. (2000). Mindful learning. *Current Directions in Psychological Science*, 9, 220-223.

Langer, E. J. & Piper, A. J. (1987). The prevention of mindlessness. *Journal of Personality and Social Psychology*, 53, 280-287.

Lane, R. D., Nadel, L. & Kaszniak, A. W. (2000). The future of emotion research from the perspective of cognitive neuroscience. In R. D. Lane & L. Nadel (Eds.), *Cognitive neuroscience of emotion* (pp. 407-410). New York, NY: Oxford University Press.

LeDoux, J. (2002). *Synaptic Self: How our brains become who we are.* New York, NY: Russell Sage Foundation.

LeDoux, J. (1996). *The emotional brain.* New York, NY: Simon & Shuster.

Lee, D. & Seo, H. (2007). Mechanisms of reinforcement learning and decision making in the primate dorsolateral prefrontal cortex. In B. W. Balleine, K. Doya, J. O'Doherty & M. Sakagami (Eds.). Reward and decision making in corticobasal ganglia networks. *Annals of the New York Academy of Sciences*, 1104, 108-122.

Lichter, D. G. & Cummings, J. L. (2001). Introduction and overview. In D. G. Lichter & J. L. Cummings (Eds.), *Frontal-Subcortical circuits in psychiatric and neurological disorders* (1-43). New York, NY: Guilford Press.

Lieberman, P. (2006). *Toward an evolutionary biology of language.* Cambridge, MA: Harvard University Press.

Lim, M. M., Wang, Z., Olazabal, D. E., Ren, X., Terwilliger, E. F. & Young, L. J. (2004). Enhanced partner preference in a promiscuous species by manipulating the expression of a single gene. *Nature,* 429, 754-757.

Lindley, D. (2007). *Uncertainty: Einstein, Heisenberg, Bohr, and the struggle for the soul of science.* New York, NY: Doubleday.

Logothetis, N. K. (2004). Higher cognitive functions. In M. S. Gazzaniga (Ed.), *The cognitive neurosciences* III (pp. 849-969). Cambridge, MA: MIT Press.

Long, D. L., Baynes, K. & Prat, C. (2007). Sentence discourse representation in two cerebral hemispheres. In F. Schmalhofer & C. A. Perfetti (Eds.), *Higher level language process in the brain: Inference and comprehension processes* (pp. 329-353). Mahwah, NJ: Lawrence Erlbaum Associates Inc.

Lorenz, K. (1965). *Evolution and modification of behavior.* Chicago, IL: University of Chicago Press.

Luria, A. R. (1981). *Language and cognition.* J. V. Wertsch (Ed.).Washington, DC: John Wiley & Sons Inc.

Luria, A. R. (1966/1980). *Higher cortical functions in man.* New York, NY: Basic Books.

MacLean, P. D. (1990). *The triune brain in evolution.* New York, NY: Plenum Press.

Malloy, P. F. & Richardson, E. D. (2001). Assessment of frontal lobe functioning. In P. S. Salloway, P. F. Malloy & J. D. Duffy (Eds.), *The frontal lobes and neuropsychiatric illness* (pp. 125-137). Washington, DC: American Psychiatric Publishing.

McInerny, D. Q. (2005). *Being logical: A guide to good thinking.* New York, NY: Random House.

Mega, M. S. & Cummings, J. L. (2005). Frontal subcortical circuits: Anatomy and function. In P. S. Salloway, P. F. Malloy & J. D. Duffy (Eds.), *The frontal lobes and neuropsychiatric illness* (pp. 15-32). Washington, DC: American Psychiatric Publishing.

Mesulam, M.-Marsel (2000). *Principles of behavioral and cognitive neurology* (2nd Ed.). New York, NY: Oxford University Press.

Mesulam, M.-Marsel (2002). The human frontal lobes: Transcending the default mode through contingent coding. In D. T. Stuss & R. T. Knight (Eds.),

Principles of frontal lobe function (pp. 8-30). New York, NY: Oxford University Press.
Mesulam, M.-Marsel (1985). Patterns in behavioral neuroanatomy: Association areas, the limbic system, and hemispheric specialization. In M. Mesulam (Ed.) *Principles of Behavioral Neurology* (pp. 1-70). Philadelphia, PA: F. A Davis.
Middleton, F. A. & Strick, P. L. (2001). A revised neuroanatomy of frontal-subcortical circuits. In D. G. Lichter & J. L. Cummings (Eds.), *Frontal-Subcortical circuits in psychiatric and neurological disorders* (44-58). New York, NY: Guilford Press.
Milgram, S. (2004). *Obedience to authority* (first published 1974). New York, NY: HarperCollins.
Milner, B, & Petrides, M. (1984). Behavioral effects of frontal-lobe lesions in man. *Trends in Neuroscience*, 7:403-407.
Minkowski, H. (1952). Space and Time. In H. A. Lorentz, A. Einstein, H. Minkowski, & H. Weyl, *The principle of relativity: A collection of original memoirs on the special and general theory of relativity*. New York, NY: Dover.
Miresco, M. J. & Kirmayer, L. J. (2006). The persistence of mind-brain dualism in psychiatric reasoning about clinical scenarios. *American Journal of Psychiatry,* 163, 913-918.
Mitchell, J. P., Mason, M. F., Macrae, C. N. & Mahzarin, R. B. (2006). Thinking about others: The neural substrates of social cognition. In J. T. Cacioppo, P. S. Visser & C. L. Pickett (Eds.), *Social neuroscience, people thinking about people* (pp. 63-82). Cambridge, MA: MIT Press.
Mitchell, J. P., Macrae, C. N. & Banaji, M. R. (2006). Dissociable medial prefrontal contributions to judgments of similar and dissimilar others. *Neuron*, 50, 655-663.
Morris, J. & Dolan, R. (2004). Emotion and memory: Functional neuroanatomy of human emotion. In R. S. J. Frackowiak, K. J. Friston, C. D. Frith, R. J. Dolan, C. J. Price, S. Zeki, J. Ashburner & W. Penny (Eds.), *Human Brain Function*, 2nd Ed., (pp. 365-396), San Diego, CA: Elsevier Science USA.
Nair, H. P. & Young, L. J. (2006). Vasopressin and pair-bond formation: Genes to brain and behavior. *Physiology*, 21, 145-152.
Neisser, U. (1967). *Cognitive psychology*. New York, NY: Appleton-Century-Crofts.
Niv, Y. (2007). Cost, benefit, tonic, phasic: What do response rates tell us about dopamine and motivation? In B. W. Balleine, K. Doya, J. O'Doherty & M.

Sakagami (Eds.). *Reward and decision making in corticobasal ganglia networks. Annals of the New York Academy of Sciences*, 1104, 357-76.
Norris, C. J. & Cacioppo, J. T. (2007). I know how you feel: Social and emotional information processing in the brain. In E. Harmon-Jones & P. Winkielman (Eds.), *Social neuroscience: Integrating biological explanations of social behavior* (pp. 84-105). New York, NY: Guilford Press.
Nowak, M. A. (2006). *Evolutionary dynamics: Exploring the equations of life.* Cambridge, MA: Belknap Harvard University Press.
Ochsner, K. N. (2007). How thinking controls feelings: A cognitive neuroscience approach. In E. Harmon-Jones & P. Winkielman (Eds.), *Social neuroscience: Integrating biological explanations of social behavior* (pp. 106-133). New York, NY: Guilford Press.
Ochsner, K. N. (2006). Characterizing the functional architecture of affect regulation: Emerging answers and outstanding questions. In J. T. Cacioppo, P. S. Visser & C. L. Pickett (Eds.), *Social neuroscience, people thinking about people* (pp. 245-268). Cambridge, MA: MIT Press.
Ochsner, K. N., Bunge, S. A., Gross, J. J. & Gabrieli, J. D. E. (2005). Rethinking feelings: An fMRI study of the cognitive regulation of emotion. In J. T. Cacioppo & G. G. Berntson (Eds.), *Social neuroscience: Key readings in social psychology* (pp. 253-270). New York, NY: Psychology Press.
O'Doherty, J. P., Hampton, A., & Kim, (H. 2007). Model-Based fMRI and its application to reward learning. In B. W. Balleine, K. Doya, J. O'Doherty & M. Sakagami (Eds.). *Reward and decision making in corticobasal ganglia networks. Annals of the New York Academy of Sciences*, 1104, 35-53.
Ogar, J. & Gorno-Tempini, M. L. (2007). The orbitofrontal cortex and the insula. In B. L. Miller & J. L. Cummings (Eds.), *The human frontal lobes*, 2nd Ed. (pp. 59-67). New York, NY: Guilford Press.
Padoa-Schioppa, C. & Assad, J. A. (2006). Neurons in the orbitofrontal cortex encode economic value. *Nature*, 441, 223-226.
Pandya, D. N., & Barnes, C. L. (1987). Architecture and connections of the frontal lobe. In E. Perecman (Ed.), *The frontal lobes revisited* (pp. 41-72) New York, NY: IRBN Press.
Panksepp, J. (1998). *Affective Neuroscience: The foundations of human and animal emotions.* New York, NY: Oxford University Press.
Petrides, M. (2005). Lateral and prefrontal cortex: Architectonic and functional organization. *Philosophical Transactions of the Royal Society*, 360, 781-795.
Petrides, M. & Pandya, D. N. (2002). Association pathways of the prefrontal cortex and functional observations. In D. T. Stuss & R. T. Knight (Eds.),

Principles of frontal lobe function (pp. 31-50). New York, NY: Oxford University Press.

Phelps, E. A. & LaBar, K. S. (2006). Functional neuroimaging of emotion and social cognition. In R. Cabeza & A. Kingstone (Eds.), *Handbook of functional neuroimaging of cognition* (pp. 421-453). Cambridge, MA: MIT Press.

Phelps, E. A. (2004). The human amygdala and awareness: Interactions between emotion and cognition. In M. S. Gazzaniga (Ed.), *The cognitive neurosciences* III (pp. 1005-1015). Cambridge, MA: MIT Press.

Poldrack, R. A. & Willingham, D. T. (2006). Functional neuroimaging of skill learning. In R. Cabeza & A. Kingstone (Eds.), *Handbook of functional neuroimaging of cognition* (pp. 114-148). Cambridge, MA: MIT Press.

Portas, C., Maquet, P., Rees, G., Blakemore, S. & Frith, C. (2004). Higher cognitive functions: The neural correlates of consciousness. In R. S. J. Frackowiak, K. J. Friston, C. D. Frith, R. J. Dolan, C. J. Price, S. Zeki, J. Ashburner & W. Penny (Eds.), *Human Brain Function* (pp. 269-301) (2nd Ed). San Diego, CA: Elsevier Science USA.

Preuschoff, K. & Bossaerts, P. (2007). Adding prediction risk to the theory of reward learning. In B. W. Balleine, K. Doya, J. O'Doherty & M. Sakagami (Eds.). Reward and decision making in corticobasal ganglia networks. *Annals of the New York Academy of Sciences*, 1104, 135-46.

Raichle, M. E. (2006). Functional neuroimaging: A historical and physiological perspective. In R. Cabeza & A. Kingstone (Eds.), *Handbook of functional neuroimaging of cognition* (pp. 3-20). Cambridge, MA: MIT Press.

Random House: *Webster's College Dictionary* (1997). New York, NK: Random House Inc.

Risberg, J. (2006). Evolutionary aspects on the frontal lobes. In J. Risberg & J. Grafman (Eds.), *The frontal lobes, development, function and pathology* (pp. 1-20). New York, NY: Cambridge University Press.

Roepstorf, A. (2004). Mapping brain mappers: An ethnographic coda. In R. S. J. Frackowiak, K. J. Friston, C. D. Frith, R. J. Dolan, C. J. Price, S. Zeki, J. Ashburner & W. Penny (Eds.), *Human Brain Function* (pp. 1105-17), 2nd Ed. San Diego, CA: Elsevier Science USA.

Rolls, T. R. (2002). The functions of the orbitofrontal cortex. In D. T. Stuss & R. T. Knight (Eds.), *Principles of frontal lobe function* (pp. 354-375). New York, NY: Oxford University Press.

Salloway, S. P. & Blitz, A. (2002). Introduction to functional neural circuitry. In G. B. Kaplan & R. P. Hammer (Eds.) *Brain circuitry and signaling in psychiatry: Basic science and clinical implications* (pp. 1-29). Washington, DC: American Psychiatric Publishing Inc.

Sapir, E. (1949). *Selected writings of Edward Sapir*. Berkeley: University of California Press.

Schlacter, D. (2001). *The Seven Sins of Memory*. New York, NY: Houghton Mifflin.

Schultz, W. & Tremblay, L. (2006). Involvement of primate orbitofrontal neurons in reward, uncertainty, and learning. In D. H. Zald & S. L. Rauch (Eds.), *The orbital frontal cortex* (pp. 173-198). New York, NY: Oxford University Press.

Schmalhofer, F. & Perfetti, C. A. (2007). Neural and behavioral indicators of integration processes across sentence boundaries. In F. Schmalhofer & C. A. Perfetti (Eds.), *Higher level language process in the brain: Inference and comprehension processes* (pp. 161-188). Mahwah, NJ: Lawrence Erlbaum Associates Inc.

Shiffrin, & Atkinson, (1969). Storage and retrieval processes in long-term memory. *Psychological Review,* 76, 179-193.

Shilpa, P., Rodak, K. L., Manikonyan, E., Singh, K. & Platek, S. M. (2007). Introduction to evolutionary cognitive neuroscience methods. In S. M. Platek, J. P. Keenan & T. K. Shackelford (Eds.), *Evolutionary cognitive neuroscience* (pp. 47-62). Cambridge, MA: MIT Press.

Siegel, D. J. (1999). *The developing mind.* New York, NY: Guilford Press.

Skinner, B. F. (1953). *Science and human behavior*. New York, NY: Macmillan Publishing Inc.

Striedter, G. F. (2005). *Principles of brain evolution*. Sunderland, MA: Sinauer Associates

Stuss, D. T., Terence, P. W. & Alexander, M. P. (2001). Consciousness, self-awareness, and the frontal lobes. In P. S. Salloway, P. F. Malloy & J. D. Duffy (Eds.), *The frontal lobes and neuropsychiatric illness* (pp. 101-109). Washington, DC: American Psychiatric Publishing.

Tapiero, I. & Fillon, F. (2007). Hemispheric asymmetry in the processing of negative and positive emotional inferences. In F. Schmalhofer & C. A. Perfetti (Eds.), *Higher level language process in the brain: Inference and comprehension processes* (pp. 355-377). Mahwah, NJ: Lawrence Erlbaum Associates Inc.

Thompson-Schill, S. L., Kan, I. P. & Oliver, R. T. (2006). Functional neuroimaging of semantic memory. In R. Cabeza & A. Kingstone (Eds.), *Handbook of functional neuroimaging of cognition* (pp. 149-190). Cambridge, MA: MIT Press.

Toates, F. M. (2007). *Biological Psychology* (2nd Ed). Harlow, UK: Pearson Education Limited.

Tomasello, M. (2004). What kind of evidence could refute the UG hypothesis? *Studies in Language,* 28, 642-645.

Tomasello, M. (2005). *Constructing a language: A usage-based theory of language acquisition.* Cambridge, MA: Harvard University Press.

Tom, S. M., Fox, C. R., Trepel, C. & Poldrack, R. A. (2007). The neural basis of loss aversion in decision-making under risk. *Science,* 315, 515-518.

Tranel, D. (2000). Electrodermal activity in cognitive neuroscience: Neuroanatomical and neuropsychological correlates. In R. D. Lane & L. Nadel (Eds.), *Cognitive neuroscience of emotion* (p. 218). New York, NY: Oxford University Press.

Tranel, D. (2002). Emotion, decision making, and the ventromedial prefrontal cortex. In D. T. Stuss & R. T. Knight (Eds.), *Principles of frontal lobe function* (pp. 338-353). New York, NY: Oxford University Press.

Tse, D., Langston, R. F., Kakeyama, M., Bethus, I., Spooner, P., Wood, E., Witter, M. & Morris, R. (2007). Schemas and memory consolidation. *Science,* 316, 76-82.

Tversky, A. & Kahneman, D. (1992). Advances in prospect theory: Cumulative representation of uncertainty. *Journal of Risk & Uncertainty,* 5, 297-323.

Vygotsky, L. S. & Luria, A. R. (1993). *Studies on the history of behavior: Ape, primitive, and child.* V. I. Golod & J. E. Knox (Editors & Translators). Mahwah, NJ: Lawrence Erlbaum Associates Inc.

Wagner, A. D., Bunge, S. A. & Badre, D. (2004). Cognitive control, semantic memory, and priming: Contributions from the prefrontal cortex. In M. S. Gazzaniga (Ed.), *The cognitive neurosciences* III (pp. 709-725). Cambridge, MA: MIT Press.

Whorf, B. L. (1956). *Language, thought and reality.* Cambridge, MA: MIT Press.

Wingfield, J. C., More, I. T., Goyman, W., Wacker, D. W. & Sperry, T. (2006). Context and ethology of vertebrate aggression: Implications for the evolution of hormone-behavior interactions. In R. J. Nelson (Ed.), *Biology of Aggression* (pp. 179-210). New York, NY: Oxford University Press.

Wright, J. H. (2004). In J. H. Wright (Ed.), *Cognitive behavioral therapy.* Washington, DC: American Psychiatric Publishing.

Young, L. J. & Wang, Z. (2004). The neurobiology of pair bonding. *Nature Neuroscience,* 7, 1048-1054.

INDEX

4

4-hydroxynonenal, 57

A

aberrant, 8
abnormalities, 5, 34, 39, 70
ACC, 103, 104
access, 78
accountability, 121
accounting, 24
accuracy, viii, 55, 85, 86, 90, 91, 92, 93, 94, 96, 101, 103, 106, 108, 109, 110, 112, 113, 116, 118, 119, 120, 121, 122, 123, 124, 126, 127, 128, 129, 131
achievement, 39, 124, 127
acidic, vii, 2, 41
activation, 7, 26, 38, 48, 55, 56, 57, 63, 78, 79, 81, 83, 105
active site, 4
Adams, 27, 78
adaptability, 99, 119, 121, 127
adaptation, 42, 47, 87, 94, 95, 99, 101, 105, 125
addiction, viii, 21
adenosine, 22
adenosine triphosphate, 22
adhesion, 4, 41

adjustment, 118
administration, 24, 71
adolescence, 35, 36, 37, 38, 39, 44, 60, 62, 68, 70, 78
adolescents, 82, 111
adult, 38, 39, 40, 45, 46, 48, 49, 60, 61, 66, 68, 69, 72, 73, 75, 76, 80, 81, 82, 99, 107, 114, 119, 129
adulthood, 29, 35, 36, 37, 44, 57, 60, 62, 63, 68, 75, 76, 111, 112, 113, 133
adults, 37, 49, 62, 66, 81, 82, 83, 97, 98, 112, 114, 121, 129
age, 36, 37, 38, 42, 43, 44, 45, 46, 47, 49, 50, 51, 53, 54, 57, 58, 60, 62, 65, 66, 70, 74, 77, 78, 82, 100, 111, 129
ageing, 68
agent, 52
aggregates, vii, 1
aggression, 88, 100, 128, 133, 141
aggressive behavior, 90
aging, 42, 44, 48, 49, 52, 62, 64, 65, 66, 67, 68, 69, 74, 76, 77, 81, 82
air, 39, 98, 133
Albert Einstein, 122
allele, 17, 55, 76, 82, 83
alleles, 77
alpha, 15, 41, 46, 62, 72
alpha-2-macroglobulin, 72
alternative, 2, 48, 113
alternatives, 98, 113
alters, 15

Alzheimer, v, vii, viii, 1, 2, 12, 13, 14, 15, 16, 17, 18, 19, 33, 34, 46, 50, 51, 62, 63, 64, 65, 66, 67, 68, 69, 70, 71, 72, 73, 74, 75, 76, 77, 78, 79, 80, 81, 82, 83
Alzheimer disease, 17, 19, 62, 63, 64, 67, 68, 69, 70, 73, 74, 77, 80, 81
ambiguity, 113
American Psychiatric Association, 130
amine, 24, 29, 30
amines, 23
amino, 4, 6, 17
amino acid, 4, 6, 17
amphetamine, 24, 28, 30, 31, 48, 62, 76, 82
amputation, 48
amygdala, 88, 104, 105, 109, 124, 139
amyloid, vii, 1, 2, 11, 12, 13, 14, 15, 16, 17, 18, 19, 41, 50, 52, 55, 57, 66, 68, 69, 71, 72, 74, 77, 78, 82, 83
Amyloid, v, 1, 46, 52, 55, 73, 74, 78
amyloid beta, 12, 13, 15, 16, 17, 18, 19, 52, 78
amyloid fibrils, 52
amyloid plaques, vii, 1, 2, 50
amyloid precursor protein, vii, 1, 2, 11, 12, 13, 14, 15, 16, 17, 18, 19, 41, 68, 69, 71, 72, 75, 82, 83
amyloidosis, 67
amyotrophic lateral sclerosis, 7, 18
anatomy, 35, 64, 79, 133
anger, 88, 124, 131
animal models, 76
animal studies, 40
animals, 24, 88, 94
anisotropy, 78
anterior cingulate cortex, 103
anthropology, vii, 128
anti-apoptotic, 25, 26
antibody, 75
antigen, 64
anti-oxidant, 7
anxiety, 88, 113, 116, 118, 125
aphasia, 75, 82
APOE, 55, 77, 82, 83
Apolipoprotein E, 17

apoptosis, 22, 25
apoptotic, 25
APP, vii, 1, 2, 3, 4, 5, 6, 7, 8, 9, 10, 11, 13, 14, 15, 17, 46, 49, 52, 55, 60, 61, 80
application, 80, 86, 111, 127, 134, 138
appraisals, 90, 93, 103, 104, 106
arginine, 87
argument, 34, 59, 116, 130
Aristotelian, 135
Army, 58
artery, 75
artificial, vii, 7
assessment, 56, 58, 83, 110, 120
assignment, 109
associations, 87, 103, 107, 109, 128
assumptions, 92, 98, 99, 109, 112, 114, 116, 117, 119, 120
astrocytes, 23
asymmetry, 140
asymptomatic, 34, 50, 52, 54, 56, 79, 80
Atlas, 134
atrophy, 39, 44, 54, 69, 74, 79, 82
attachment, 87, 88, 133, 135
attention, 61, 90, 104, 106, 133, 134
atypical, 28, 132
authority, 113, 116, 117, 122, 123, 137
autobiographical memory, 103
autonomic, 104
autonomy, 110
autopsy, 51, 52
autosomal dominant, 46
availability, 34, 109
aversion, 141
awareness, 86, 92, 94, 98, 103, 111, 114, 115, 116, 120, 121, 127, 139
axonal, 5, 6, 14, 35, 40
axonal degeneration, 5
axons, 6, 45

B

basal ganglia, 105, 130, 134
Bax, 25, 26
BDNF, 26, 88

behavior, viii, 30, 35, 45, 60, 85, 86, 87, 88, 89, 90, 91, 92, 102, 103, 105, 106, 107, 112, 114, 118, 119, 121, 123, 125, 126, 127, 128, 129, 133, 134, 136, 137, 141
behavior therapy, 133
behaviours, 24
belief systems, viii, 86, 88, 89, 90, 91, 94, 98, 99, 100, 101, 106, 109, 111, 112, 115, 117, 118, 119, 122, 125, 128, 129
beliefs, 89, 90, 91, 92, 93, 96, 98, 100, 103, 106, 112, 113, 114, 117, 118, 120, 123, 124, 125, 126, 128, 129
benefits, 114, 121, 126
bereavement, 88
beta, 3, 12, 13, 14, 15, 16, 17, 18, 19, 41, 52, 57, 63, 64, 66, 72, 74, 77, 83
bias, viii, 85, 91, 92, 96, 97, 98, 101, 103, 107, 108, 110, 111, 112, 114, 116, 118, 119, 121, 124
binding, 7, 14, 41, 46, 117
biochemical, 2, 13, 46, 81
biochemistry, 133
biologic, 7, 39, 45, 49, 60
biological, 2, 4, 12, 95, 110, 132, 138
biologically, 39, 46, 100
biology, vii, 19, 129, 136
biomarker, 57
biomarkers, 57, 59, 67
biometry, 71
biopsy, 51
birth, 5, 35, 36, 133
black, 92, 99, 102, 122, 124, 125
blame, 124, 125
blaming, 98, 123, 124, 125
blocks, 7, 117
Bohr, 136
bonding, 87, 88, 89, 90, 141
bottom-up, 107, 108, 111
boutons, 45, 66, 80
boys, 36, 78
brain activity, 66
brain development, 35, 39, 41, 44, 65, 69
brain functioning, viii, 85, 90
brain growth, 44, 80

brain imaging techniques, 36
brain injury, 48, 69, 76, 81
brain size, 37, 44, 64, 76
brain structure, viii, 85, 110, 126, 128
brainstem, 24, 37
building blocks, 97
bullying, 126
bundling, 5
business, 131

C

Ca^{2+}, 47, 71
calcium, 41, 48
California, 140
cAMP, 82
Canada, 131
candidates, 99
capacity, 37, 48, 49, 90, 108, 110, 116, 122
carboxyl, 16
caregivers, 96
cargo, 7, 9
carrier, 6, 57
casein, 6
caspase, 25
caspase-dependent, 25
cast, 123
catabolism, 24
catalytic, 3, 5, 26
catalytic activity, 5
catecholamine, 24, 27
categorization, 94, 95
cation, 16
Caucasian, 71
Caucasian population, 71
causal relationship, 120
causality, 94, 97
causation, 125
cell, viii, 2, 6, 8, 10, 12, 15, 19, 21, 22, 23, 24, 25, 26, 28, 30, 64, 76, 100
cell adhesion, 15
cell culture, 8
cell cycle, 25, 30, 64
cell death, viii, 21, 25, 26, 28, 30

cell line, 22
cell lines, 22
cell surface, 2, 6, 8, 10
central executive, 131, 134
central nervous system, 5, 23, 29, 78
cerebral amyloidosis, 72
cerebral cortex, viii, 21, 22, 23, 25, 26, 28, 30, 62, 63, 68, 82
cerebral hemisphere, 37, 136
cerebrospinal fluid, 43, 75
cerebrum, 70
certainty, 92, 113, 119, 124
changing environment, 95
channels, 4, 19
Charles Darwin, 121
Chicago, 136
childhood, 37, 58, 60, 62, 67, 68, 70, 76, 78, 81, 107, 111, 123, 129
childhood disorders, 81
children, 37, 62, 70, 78, 81, 95, 97, 98, 100, 112, 129, 130, 131
Chinese, 22
chromosome, 3
chronic, 65
classes, 123
classification, 62, 91
classified, vii
cleavage, vii, 1, 2, 3, 6, 7, 8, 12, 13, 15, 17, 18, 52
cleavages, 2, 8
clinical, 27, 34, 39, 50, 51, 52, 53, 54, 58, 59, 60, 61, 67, 69, 73, 74, 75, 80, 132, 137, 139
clinical approach, 50
clinical assessment, 58
clinical diagnosis, 51
clinical symptoms, 27
clinical trial, 52, 53, 59
clinical trials, 52, 59
cloning, 17, 22, 28
CNS, 5, 28, 37, 72
Co, 132
cocaine, 31
coding, 106, 136
codon, 16

cognition, vii, viii, 45, 54, 60, 71, 78, 85, 86, 87, 88, 90, 93, 99, 100, 102, 103, 104, 105, 106, 107, 108, 115, 118, 120, 121, 122, 125, 126, 127, 128, 129, 131, 132, 133, 134, 136, 139, 140
cognitive activity, 65
cognitive disorders, 81
cognitive domains, 58
cognitive function, 42, 44, 50, 80, 105, 115, 127, 133, 136, 139
cognitive impairment, 40, 59, 63, 65, 67, 68, 73, 75
cognitive performance, 35, 40, 57, 60, 76
cognitive process, viii, 85, 90, 92, 93, 105, 106, 114
cognitive processing, 90, 93, 105
cognitive science, vii
cognitive system, 132
cognitive test, 57
cognitive testing, 57
coherence, 116
collateral, 92, 119
colors, 38
commissure, 87
communication, 87, 88, 89, 92, 93, 94, 96, 114, 117, 119, 125
community, 39, 63, 78
community-based, 63
compensation, 45, 48, 49, 60
competence, 97
competency, 108
competition, 7, 100, 115
complement, 37, 98
complementary, 46, 59, 99
complex systems, 100
complexity, 41, 47, 58, 62, 66, 99, 100, 103, 107, 109, 121, 122
components, 2, 22, 41, 44, 47, 54, 86, 93, 94, 104, 109, 115, 118
comprehension, 133, 136, 140
computer, vii, 102, 103, 108
computer science, vii
computer software, 102
computers, 102
concentration, 7

conception, 35
conceptualization, 96
conceptualizations, 96
concrete, 91, 92, 95, 110, 112, 114
Concrete, 111, 112
concreteness, 99, 100, 118
conditioning, 101
confabulation, 126
confidence, 100
confirmation bias, 123
conflict, 59, 116
conflict resolution, 116
confrontation, 118
confusion, 86
Congress, iv
congruence, 103
conjecture, 95
connectionist, vii
connectivity, 70, 109, 111
consciousness, 139
consensus, 118
conservation, 93
consolidation, 36, 42, 141
constraints, 110, 117
context-dependent, 127
context-dependent manner, 127
contingency, 93, 94, 108
continuing, 96
continuity, 36
control, viii, 21, 25, 44, 64, 105, 112, 125, 133, 134, 135, 141
controlled, 72, 82, 99, 108
conversion, 54, 57, 64, 69
coordination, 87
copper, vii, 2, 7, 12
core-shell, 134
corpus callosum, 37, 43, 67, 76
correlation, 55, 70, 98
correlations, 99
cortex, 25, 26, 40, 41, 45, 48, 51, 66, 69, 80, 81, 82, 102, 105, 130, 132, 135
cortical, 5, 25, 26, 28, 35, 36, 37, 38, 40, 41, 44, 45, 48, 49, 62, 63, 64, 66, 67, 68, 71, 74, 76, 105, 108, 111, 136
cortical functions, 136

cortical inhibition, 49
cortical neurons, 25, 26, 68
corticospinal, 37
coupling, 47, 106
CREB, 46, 47, 72, 81
critical points, 41
critical thinking, 96, 116, 121, 127, 129, 131, 132
critical thinking skills, 96
critical variables, 39
cross-cultural, 93
cross-sectional, 43, 64, 76, 78, 81
cross-sectional study, 43, 64, 81
CSF, 43, 57, 59
C-terminal, vii, 1, 2
cues, 107, 109, 110
cultural, viii, 34, 39, 85, 86, 87, 88, 89, 90, 91, 93, 94, 95, 96, 97, 98, 99, 100, 101, 106, 107, 109, 110, 111, 112, 113, 114, 116, 117, 118, 119, 122, 123, 124, 125, 127, 128, 129
cultural artifacts, 98
cultural beliefs, viii, 85, 89, 91, 93, 101, 106, 109, 113, 118, 124, 128, 129
cultural factors, 34
cultural influence, 39
cultural stereotypes, 123
culture, 89, 91, 93, 95, 97, 98, 99, 101, 111, 113, 116, 118, 128
cycles, 83
cyclic AMP, 46
cysteine, 41
cytoarchitecture, 35
cytoplasm, 22, 27
cytoplasmic tail, 3
cytosolic, 7, 12, 14, 17

D

daily living, 50
danger, 113
death, 25, 26, 28, 29, 31, 83
decay, 109
deception, 132

decision making, 90, 91, 106, 114, 122, 132, 135, 138, 139, 141
decision-making process, 111
decisions, 90, 98, 103, 104, 105, 108, 109, 110, 111, 112, 115, 117, 120, 124, 126, 127, 129
deduction, 127
defects, 58
defensiveness, 100
deficit, 80
deficits, 74
definition, 36, 50, 92, 118
degradation, 6, 24, 26
degrading, 4
degree, 42, 50, 91, 115, 116, 119, 124, 127
degrees of freedom, 110
delays, 60
delta, 46, 69
delusion, 118
delusions, 118
demand, 98
dementia, viii, 33, 34, 35, 39, 40, 44, 49, 50, 51, 52, 53, 54, 55, 58, 59, 61, 67, 68, 70, 71, 73, 77, 78, 82, 83
demyelination, 71
dendritic spines, 44, 45, 66, 78
density, 4, 18, 35, 36, 37, 38, 42, 44, 55, 58, 72, 78, 80
Department of Health and Human Services, 73
depolarization, 24
deposition, 15, 65
deposits, vii, 1
depression, 45, 88, 116
derivatives, 29
desire, 98
destruction, 63
detachment, 88
detection, viii, 33, 34, 50, 55, 58, 59, 61, 74, 94, 96, 103
developing brain, 28, 83
developmental change, 38
developmental factors, 60
developmental process, 44
deviation, 113, 125

diagnostic, 50, 51, 73, 118
Diagnostic and Statistical Manual of Mental Disorders, 130
Diamond, 115, 116, 133
dichotomy, 134
diet, 39
differentiation, 23, 73
diffusion, 78, 83
diffusivity, 35, 37, 78
dihydroxyphenylalanine, 27
dimerization, 7
discomfort, 113
discontinuity, 36
discordance, 39, 125
discourse, 136
discriminatory, 97, 112
disease progression, 69
diseases, 72
disorder, vii, 1, 27
dissociation, 69
distortions, 64, 116
distribution, 7, 16, 36, 38, 51, 68, 82, 87
disulfide, 5, 7
disulfide bonds, 5
divergent thinking, 118
diversity, 100, 119
division, 41
dominance, 98, 99, 100, 122, 125
dopamine, viii, 21, 22, 23, 26, 28, 87, 137
dopaminergic, 23, 29, 48
dopaminergic neurons, 23, 29
dorsolateral prefrontal cortex, 102, 135
Down syndrome, 69
drug exposure, 39
drugs, 48
D-serine, 74
dualism, 92, 117, 137
dynamic environment, 95, 105, 119, 123
dynamic systems, vii
dynamical system, vii
dysregulation, 48

E

eating, 118

eating disorders, 118
ecological, 93
economic, 128, 138
economics, 132
education, 35, 39, 40, 44, 50, 58, 62, 64, 71, 73, 76, 80, 82, 91, 111, 125, 129
Education, 39, 70, 77, 140
efficacy, 27, 109
EGF, 83
Einstein, 136, 137
elaboration, 35, 41
elderly, 50, 65, 66, 67, 69, 70, 71, 73, 74, 76, 79
elders, 112
electrochemical, 22
electronic, iv
electrophysiological, 5
electrostatic, iv
embryonic, viii, 21, 22, 23
embryos, 72
emission, 77
emotion, viii, 86, 88, 91, 103, 115, 120, 121, 128, 132, 135, 137, 138, 139, 141
emotional, 15, 86, 87, 88, 90, 93, 103, 104, 106, 107, 108, 109, 115, 120, 129, 131, 134, 135, 138, 140
emotional disorder, 131
emotional experience, 134
emotional information, 138
emotional reactions, 109, 120
emotional responses, 108
emotional state, 109
emotional valence, 87, 106, 107, 108, 109
emotions, 89, 90, 91, 94, 98, 102, 105, 106, 108, 109, 110, 114, 115, 116, 117, 120, 121, 124, 128, 138
empathy, 103, 105, 132
encephalopathy, 34
encoding, 41, 57, 80, 81, 109
encouragement, 130
endocytosis, 6, 14, 15
endogenous, 8, 27, 87
endoplasmic reticulum, 2, 18
engagement, 135
engineering, 76

English, 97
enlargement, 35
enterochromaffin cells, 22
entorhinal cortex, 54, 83
environment, 95, 97, 100, 103, 104, 108, 110, 114, 125
environmental, viii, 27, 47, 49, 86, 94, 102, 106, 107, 109, 110
environmental stimuli, 47, 49, 94
enzymatic, 4, 19
enzymatic activity, 19
enzyme, 2, 3, 5, 7, 12, 14, 17, 19, 24
enzymes, 3
Epictetus, 120
epigenetic, 40, 41, 42, 44, 60
epigenetic mechanism, 40, 41, 60
epilepsy, 81
epinephrine, 23
episodic, 57, 81, 109, 111
episodic memory, 57, 109
epithelial cell, 7
epithelial cells, 7
erosion, 42, 46, 49, 54, 60
error detection, 102, 103, 108, 118, 126
ethology, 128, 141
etiology, viii, 21, 29
eukaryotic, 46
evidence, 3, 42, 43, 45, 48, 53, 59, 60, 63, 69, 76, 83, 86, 92, 98, 102, 116, 118, 129, 141
evolution, viii, 59, 85, 86, 88, 94, 95, 97, 98, 99, 100, 106, 110, 126, 127, 131, 136, 140, 141
evolutionary, 22, 87, 93, 95, 99, 100, 113, 125, 128, 134, 136, 140
examinations, 53
excitability, 48
exclusion, 88
excuse, 125
execution, 105, 108
executive function, viii, 44, 78, 85, 90, 92, 104, 105, 106, 107, 109, 110, 111, 112, 115, 122, 127, 131
executive functioning, viii, 86, 90, 105, 107, 109, 110, 111, 122, 127, 131

executive functions, 44, 78, 104, 107, 112, 131
experimental design, 54
expert, iv, 51
expertise, 58
exponential, 51, 97
exposure, 45, 95, 129
Exposure, 27
external environment, 45, 104, 110
extracellular, vii, 1, 4, 24, 46
extracellular signal regulated kinases, 46

F

fabric, 106
facial expression, 96, 105, 125
failure, 24, 82, 97, 124
false, 92, 99, 116, 117, 118, 120
false belief, 116, 118
familial, 7, 12, 80, 100
family, vii, 2, 3, 13, 14, 22, 39, 41, 55, 57, 72
family history, 55, 57
family members, 14
fear, 124
feedback, 94
feeding, 24
feelings, 109, 120, 121, 138
females, 37
fetal, 28, 62, 71
F-FDG PET, 75
fiber, 37, 38, 60
fibers, 37, 44
fire, 133
flexibility, 86, 90, 94, 99, 100, 107, 108, 110, 116, 117, 121, 127, 128
float, 124
floating, 122
flow, 107
fluctuations, 71
fluid, 66, 67, 72
fluorescence, 18
fMRI, 38, 63, 66, 70, 81, 138
folding, 5, 17
food, 98

Food and Drug Administration, 52
Ford, 82
forebrain, 28, 29
formal education, 39
Fox, 70, 77, 135, 141
fractional anisotropy, 37
Framingham study, 58
France, 21
freedom, 92
Freud, 1
frontal cortex, 80, 105, 135, 140
frontal lobe, viii, 37, 42, 43, 63, 85, 90, 92, 102, 104, 105, 109, 110, 111, 112, 115, 118, 125, 130, 131, 132, 133, 134, 135, 136, 138, 139, 140, 141
frontal lobes, 42, 102, 104, 105, 109, 110, 111, 112, 115, 125, 131, 132, 133, 135, 136, 138, 139, 140
frontal-subcortical circuits, 105, 137
frontotemporal dementia, 35, 75
functional architecture, 138
functional changes, 54
functional imaging, 39, 56, 57
functional magnetic resonance imaging, 67, 78
functional MRI, 57, 132

G

gait, 78
ganglia, 132, 135, 138, 139
gauge, 124
gender, 36, 37, 79
gender effects, 79
gene, 3, 24, 26, 29, 31, 41, 47, 79, 80, 94, 120, 136
gene expression, 26
gene therapy, 79
generalization, 90, 97
generalizations, 94, 120
generation, vii, 2, 3, 4, 5, 6, 7, 8, 9, 11, 12, 13, 14, 16, 18, 37, 41, 91, 94, 95, 106, 129
genes, 17, 22, 30, 41, 46, 61, 79

genetic, 26, 28, 40, 41, 44, 62, 63, 76, 79, 81, 88, 100, 102
genetic blueprint, 102
genetics, 101
genotype, 40, 70, 74
Germany, 1
gestation, 35, 36, 40
gestational age, 36
gestures, 96, 105, 125
girls, 36, 78
glial, 4, 41, 74
glial cells, 4
gliosis, 63
glucose, 55, 75, 80
glucose metabolism, 55, 80
glutamatergic, 23, 28
glutathione, 72
glycoprotein, 4, 15
glycosylated, 5
goals, 54, 109, 123
GPI, 7, 13
grants, 11, 61
granules, 29, 30
gray matter, 36, 43, 70, 71, 76, 80
grounding, 106
groups, 7, 23, 24, 87, 89, 94, 98, 99, 114, 123
growth, 24, 35, 36, 46, 79, 88, 97
growth factor, 79, 88
guiding principles, 116
guilt, 124
guilty, 124
gut, 22
gyri, 55, 65
gyrus, 56, 103, 104

H

handling, 23
haplotypes, 29
harm, 121
harmony, 121
Harvard, 136, 138, 141
hate, 131
head, 35, 36, 37, 44, 48, 62
head injury, 48
health, 40
heart, 66
Heisenberg, 136
heterodimer, 7
heterozygote, 28, 29
heterozygotes, 30
heuristic, 111, 118
high risk, 71, 79
higher education, 39
high-level, 105
high-risk, 57
hippocampal, 5, 6, 12, 28, 35, 37, 53, 54, 62, 69, 81, 82, 88, 105
hippocampus, 25, 30, 35, 70, 78, 83, 103, 104, 109, 133
histamine, 22, 23
histogenesis, 74
histology, 35
Holland, 37, 38, 70, 78, 81, 82
homeostasis, viii, 67, 72, 79, 86, 88, 94, 95, 103, 104, 110
homogeneity, 99
homolog, 3, 13
homology, 3, 22
hormone, 141
hormones, 88, 115
hostility, 131
House, 118, 136, 139
human, viii, 7, 11, 17, 18, 28, 34, 35, 40, 41, 44, 45, 61, 62, 64, 67, 68, 69, 70, 76, 78, 79, 80, 81, 82, 85, 86, 87, 88, 89, 90, 91, 93, 94, 95, 96, 97, 99, 100, 102, 106, 107, 113, 114, 116, 117, 119, 121, 123, 124, 125, 126, 127, 128, 129, 132, 133, 134, 135, 136, 137, 138, 139, 140
human behavior, 129, 134, 140
human brain, viii, 7, 11, 17, 18, 28, 44, 62, 64, 68, 79, 85, 87, 89, 90, 91, 94, 95, 102, 116
human cerebral cortex, 76
human cognition, 90, 99, 106
human development, 61
human experience, 90

human immunodeficiency virus, 34, 80
human interactions, 99
human mental functioning, 86
humane, 127, 129
humans, viii, 85, 87, 88, 89, 91, 94, 95, 96, 98, 99, 100, 101, 102, 107, 108, 110, 111, 113, 114, 117, 119, 121, 122, 123, 125, 126, 128
Hybrid systems, vii
hydrogen, 23, 27
hydrogen peroxide, 23, 27
hypomorphic, 25
hypothalamus, 104
hypothesis, 34, 40, 44, 52, 54, 64, 68, 77, 141

I

identification, 28, 78, 93, 94, 95, 106, 110
ideology, 97
Illinois, 62
illusions, 113
image analysis, 72
imagery, 90
imaging, viii, 6, 18, 33, 53, 54, 55, 57, 61, 62, 65, 69, 72, 74, 78, 83
imaging techniques, 54
immobilization, 48
immunocytochemistry, 63
immunoreactivity, 65
impairments, 53, 71, 77
implementation, 127
in utero, 36, 44
in vitro, 16, 26, 28
in vivo, 15, 28, 65, 70, 71, 76
inactivation, 29
incidence, 80
inclusion, 57
independent variable, 87, 89, 90
indication, 52
indicators, 140
inertia, 100, 129
infancy, 75
infants, 96, 98
Infants, 96

infarction, 66, 80
infection, 80
inferences, 94, 99, 103, 116, 140
infinite, 97
information processing, 106, 110, 111, 116, 127
inheritance, 100
inherited, 11, 34, 89, 100, 101, 102, 108, 110, 114, 117, 128
inhibition, 3, 107, 114
inhibitor, 22, 23, 25, 64
inhibitors, 30
inhibitory, 135
initiation, 46
injuries, 45, 48
injury, iv, 49, 118, 130
insight, 48, 118
inspection, 93, 99
instruments, 123
insults, 116
integration, 81, 90, 103, 104, 106, 107, 110, 115, 127, 140
integrin, 62
integrins, 41
integrity, 47, 106
intelligence, vii, 35, 39
intelligence quotient, 35
intensity, 51, 53, 60
intentions, 94, 95
interaction, 4, 7, 9, 10, 18, 41, 42, 74, 104, 106, 110, 112, 131, 133
Interaction, 17
interactions, viii, 2, 7, 41, 46, 85, 87, 89, 90, 91, 96, 104, 112, 114, 121, 122, 129, 141
interdependence, 129
interdisciplinary, vii
interleukin, 4
interleukin-1, 4
Interleukin-1, 15
internalization, 3
international, 132
interpretation, 25, 70, 125
interstitial, 87
interval, 59

intervention, 11, 42, 48, 60, 69, 96
interview, 50, 67
intimidating, 98
intracranial, 44, 78
intrinsic, 35, 48, 60, 66
involution, 42, 44
ion channels, 71
irrationality, 86, 129
irritability, 100
ischemic, 62
isoforms, 57

J

JAMA, 80
Japan, 80
judge, 124
judgment, 58

K

Kentucky, 33
kidney, 6, 15
kinase, 6, 41, 46, 81
knockout, 11, 16, 27, 28, 29, 30

L

labeling, 93, 94, 95, 98, 117, 119, 123, 125
lamina, 62
laminar, 62
language, viii, 37, 38, 44, 81, 85, 86, 87, 88, 89, 90, 91, 93, 94, 95, 96, 97, 98, 99, 100, 103, 104, 106, 107, 110, 116, 125, 133, 134, 136, 140, 141
language acquisition, 141
language development, 37, 97
language lateralization, 81
language processing, 93
language skills, 110
large-scale, 58
late-onset, 17, 42, 53, 55, 59, 61
late-onset AD, 42, 53, 55, 59, 61
later life, 61

law, 51
lead, 49, 59, 91
learning, 42, 45, 46, 47, 48, 49, 56, 60, 71, 74, 81, 86, 98, 100, 102, 105, 109, 111, 112, 118, 122, 127, 135, 138, 139, 140
learning behavior, 86
left hemisphere, 38
lesions, 62, 137
Lewy bodies, 51, 73, 82
liberation, 2
lifetime, 18
ligand, 4, 15
ligands, 72
likelihood, 40, 51, 53, 54
limbic system, 87, 103, 106, 107, 109, 114, 124, 137
limitation, 58
limitations, 24, 54
linear, 35, 36
linguistic, 91, 95, 97, 100
linguistics, vii, 93
links, 105
lipase, 72
lipid, 13
Lipid, 17
lipid rafts, 13
lipoprotein, 4, 16, 18, 72
listening, 38, 70, 94, 97, 98
literature, 34, 39, 48, 50, 58
localization, 2, 9
locomotion, 30
locus, 29
locus coeruleus, 29
logical reasoning, 129
London, 76
longitudinal study, 58, 72, 77
long-term, 45, 61, 65, 78, 95, 106, 124, 140
long-term memory, 61, 106, 140
long-term potentiation, 45
losses, 42, 45, 50
love, 131

M

machinery, 25, 88
magical thinking, 112
magnetic, iv, 35, 42, 64, 71, 72, 75, 77
magnetic resonance, 35, 42, 64, 71, 72, 75, 77
magnetic resonance image, 42
magnetic resonance imaging, 35, 64, 72, 75, 77
maintenance, 23, 25, 60
maladaptive, 92
males, 37
mammalian brain, 93
mammals, 87, 88, 93, 94
management, 115
manipulation, 108
mantle, 38, 40
MAO, 23, 26
MAPK, 46, 48, 81, 82
mapping, 54, 68
marital discord, 118
maternal, 131
mathematical, 117
matrix, 24, 41, 112
maturation, 5, 36, 37, 38, 41, 62, 80, 83, 111, 129
MCI, 42, 50, 51, 52, 53, 54, 55, 56, 57, 58, 59, 61, 64, 69, 80
meanings, 95, 97
measurement, 36, 57, 78, 128
measures, 54, 55, 58, 63, 65, 69, 83, 128
mechanical, iv, 93
media, 4, 39
medial prefrontal cortex, 104
mediation, viii, 85, 86, 103
medications, 52
membership, 89
memory, 40, 46, 50, 53, 54, 55, 57, 58, 59, 65, 72, 74, 77, 79, 81, 82, 88, 90, 91, 102, 103, 104, 105, 108, 109, 110, 111, 112, 121, 137, 141
memory loss, 53
men, 75
mental ability, 58
mental health, 118
mental health professionals, 118
mental retardation, 81
mentorship, 61
metabolic, 39, 75, 79
metabolism, 9, 23, 27
metabolite, 27
metabolites, 24
metacognition, 106
methamphetamine, 27, 28
methylation, 40
mice, 3, 4, 6, 11, 13, 16, 24, 25, 26, 27, 28, 29, 30, 40, 65, 71
microscopy, 18
midbrain, 29
middle-aged, 81
migration, 38, 40, 41, 44, 68, 76
mild cognitive impairment, 34, 68, 69, 70, 72, 74, 75, 78, 79, 80, 83
Mild Cognitive Impairment (MCI), 50
military, 57
misleading, 121
misunderstanding, 86
MIT, 130, 131, 132, 134, 136, 137, 138, 139, 140, 141
mitochondrial, 25
mitogen, 41, 46
mitogen-activated protein kinase, 41, 46
mitotic, 41
mobility, 78
modality, 37, 38
modeling, 100, 109
models, vii, viii, 7, 21, 24
modulation, 67, 77, 133
molecular biology, 46
molecular mechanisms, viii, 2, 60
molecular oxygen, 23
molecular weight, 72
monoamine, viii, 21, 22, 23, 24, 28, 29, 30
monoamine oxidase, 23, 28, 30
monoaminergic, 23
monoclonal, 15, 64, 66
monoclonal antibody, 15, 64, 66
morphogenesis, 71
morphological, 46

morphology, 24, 45, 65, 70, 75
morphometric, 42, 65, 68, 79
mortality, 24
motivation, 34, 103, 104, 124, 137
motor function, 65, 94
motor neurons, 28
mouse, 4, 25, 29, 30, 72
mouse model, 4, 29
movement, 64, 72, 105
MPP, 27, 29
MPTP, 27, 28, 29, 30
MRI, 35, 37, 64, 68, 69, 70, 80, 83
mRNA, 3, 23, 25, 26, 30
multiplicity, 60
multivariate, 91, 92, 93, 99, 100, 110, 112, 119, 122, 126, 127
muscles, 64
mutant, 7, 16
mutation, 3, 13, 41, 80
mutations, 7, 41, 55, 80
myelin, 5, 44, 64
myelination, 5, 14, 19, 36, 37, 38, 40, 41, 44, 60, 111

N

naming, 56, 57, 96, 116
narratives, 94, 120
natural, 91, 100
natural environment, 91
natural selection, 100
natural selection processes, 100
negative regulatory, 46, 61
negative valence, 88, 113
neglect, 61
neocortex, 30, 37, 38, 78, 110
neonatal, 31, 62
neonates, 83
nerve, 5, 42, 48, 60
nerve cells, 5
nerve fibers, 60
nerve growth factor, 42
nerves, 6
nervous system, 45, 72, 83
network, 18, 103

neural connection, 111
neural crest, 29
neural development, 72
neural network, vii, 104, 106, 133
neural networks, vii, 104, 106, 133
neuritic plaques, 65
neuroanatomy, 71, 133, 134, 137
neurobiological, 135
neurobiology, 44, 132, 134, 141
neurochemistry, 131
neurodegeneration, 16, 49, 73, 77
neurodegenerative, vii, 1, 27, 34, 44, 48, 80
neurodegenerative disease, 27, 34, 80
neurodegenerative diseases, 27, 34
neuroeconomics, 128, 134
neuroendocrine, 22
neuroendocrine cells, 22
neurofibrillary tangles, vii, 1, 2, 50, 54, 65, 73
neurogenesis, 35, 36, 40, 80, 82
neuroimaging, 131, 132, 134, 139, 140
neurological disorder, 136, 137
neuromodulator, 74
neuronal cells, 6
neuronal density, 35
neuronal loss, 42, 44
neuronal migration, 35, 36, 40, 41, 62, 71, 77
neuronal plasticity, 40, 45, 48, 49, 60, 66
neurons, viii, 3, 5, 6, 7, 11, 12, 21, 22, 23, 24, 25, 26, 27, 28, 29, 30, 35, 38, 40, 41, 44, 52, 66, 140
neuropathological, 70, 74, 83
neuropathologies, 34, 51
neuropathology, viii, 33, 34, 35, 39, 41, 42, 49, 51, 52, 53, 54, 55, 57, 59, 60, 69, 74, 78
neuroprotection, 64
neuropsychiatry, 134
neuropsychology, 128
neuroscience, vii, viii, 85, 86, 123, 128, 129, 130, 131, 132, 134, 135, 137, 138, 140, 141
neurotoxic, 27

neurotoxic effect, 27
neurotoxicity, 27, 28
neurotoxins, 27
neurotransmission, viii, 21
neurotransmitter, 22, 25, 71
neurotransmitters, 22, 23
neurotrophic, 26, 30
New York, iii, iv, 63, 83, 130, 131, 132, 133, 134, 135, 136, 137, 138, 139, 140, 141
Nielsen, 14
NMDA, 78
noise, 105
norepinephrine, viii, 21, 22, 23, 26, 28
normal, viii, 16, 21, 24, 26, 27, 34, 36, 38, 41, 42, 43, 44, 45, 46, 47, 49, 53, 54, 55, 56, 57, 58, 59, 60, 61, 62, 63, 64, 65, 66, 68, 69, 71, 73, 74, 77, 78, 79, 80, 82, 87, 89, 102, 116, 128
normal aging, 42, 45, 61, 64, 69, 71, 73, 77, 80
normal children, 82
normalization, 26
norms, 58
North America, 71
novelty, 103, 107
N-terminal, 5, 15
nuclear, 46
nucleus, 87
nucleus accumbens, 87

O

obedience, 122
obesity, 118
objectification, 86, 97
objective criteria, 98
objectivity, 103, 125, 128
observations, 24, 37, 45, 59, 138
occlusion, 75
occupational, 39, 62
oculomotor, 105
odds ratio, 39
older adults, 63, 66, 72, 77, 78
online, 108

opioids, 87
oral, 94
orbitofrontal cortex, 103, 135, 138, 139
organization, 37, 62, 130, 132, 134, 138
orientation, 92, 115, 121
ovaries, 22
oxidation, 27, 57
oxidation products, 57
oxytocin, 87, 132

P

packaging, viii, 21
pain, 87, 88, 129
Paper, 80
paradox, 122
parallel processing, 105
parent-child, 114, 122
parents, 100, 122, 123, 129
parietal lobe, 104, 105
parietal lobes, 104, 105
Paris, 21, 27
Parkinson, viii, 21, 22, 27, 29, 73
Parkinson disease, 29
parkinsonism, 29
passive, 38
pathogenesis, viii, 2, 3, 11
pathology, 34, 39, 48, 50, 51, 52, 53, 54, 59, 60, 63, 66, 69, 131, 132, 139
Pathophysiological, 3
pathways, 7, 16, 25, 42, 46, 87, 88, 103, 104, 105, 131, 138
patients, 4, 11, 27, 39, 48, 52, 61, 62, 64, 71, 109, 118
pattern recognition, 93, 94
patterning, 57
pepsin, 3, 13
peptide, vii, 1, 12, 13, 14, 17, 18, 67, 72, 78
peptides, 2, 12, 13, 132
percentile, 36
perception, 93, 101, 102, 103, 106, 110, 113, 126
perceptions, 90, 97, 101, 103, 106, 117
performance, 54, 57, 103, 124

perfusion, 70, 80
peripheral nerve, 19
peripheral nervous system, 14, 30
personal, 90, 111, 117, 128
personal history, 90
personal responsibility, 117
personality, 118
personality characteristics, 118
perturbations, 126
PET, 55, 57
PFC, 88, 108
pharmacological, 24, 27
pharmacological treatment, 24
pharmacology, 22
phenotype, 4, 16, 23, 26, 70, 82
Philadelphia, 83, 137
philosophy, vii
phosphate, 7, 16
phosphorylation, vii, 2, 4, 6, 8, 18, 46, 65
Phosphorylation, 18
phylogeny, 95
physical activity, 40, 65, 71, 77
physical fitness, 40
physical therapy, 48
physicians, 52
physics, vii
physiological, 7, 8, 46, 74, 139
placebo, 82
planning, 90, 94, 105, 119, 127
plaque, 4, 51, 54
plaques, vii, 1, 2
plasma, 6, 23, 57, 68
plasma membrane, 6, 23
plastic, 46, 47, 48, 49, 64, 100
plasticity, 34, 40, 46, 47, 48, 49, 63, 64, 66, 71, 72, 75, 76, 82
Plato, 117
play, viii, 5, 21, 25, 27, 40, 46, 103, 111
pleiotropy, 70
polarized, 6
poor, 35, 117
population, 23, 62, 82, 118
positive reinforcement, 121
positron, 55, 75, 77
positron emission tomography, 55, 75

postmortem, 51, 65
post-stroke, 64
post-translational, vii, 2, 40
post-translational modifications, vii, 2
power, 51
pragmatic, 86
preclinical, 66, 68, 69
predicate, 97
prediction, 57, 58, 66, 73, 94, 103, 139
predictors, 67, 71, 76
preference, 41, 109, 136
prefrontal cortex, 87, 105, 107, 115, 132, 133, 134, 138, 141
prefrontal cortex (PFC), 87
prejudice, 123
premature infant, 83
preparation, iv
presenilin 1, 80
presynaptic, 6, 66, 81
prevention, 34, 135
preventive, 59
primary school, 39
primate, 79, 88, 107, 127, 135, 140
primates, 42, 87, 88, 90, 93, 107, 131
priming, 141
priorities, 105
proactive, 92, 121
probability, 86, 92, 94, 99, 107, 109, 113, 115, 119, 121, 125, 127
problem solving, 90, 91, 95, 111, 115, 119, 135
production, vii, 2, 3, 6, 11, 12, 15, 16, 52, 63, 64
progenitor cells, 74
program, 61, 72
programming, 134
progressive, 38, 59, 75
proliferation, 41
promote, viii, 49, 85, 86, 92, 113, 129
promoter, 29
property, iv
proteases, 3, 11
protein, vii, 1, 2, 4, 9, 11, 12, 13, 15, 16, 18, 19, 23, 41, 46, 47, 49, 57, 69, 72, 73, 78

protein binding, 9
protein synthesis, 47
proteins, vii, 2, 3, 4, 5, 6, 7, 8, 9, 10, 11, 12, 14, 17, 18, 22, 41, 46, 47, 49, 69, 81, 132
proteolysis, 17
proxy, 36
pruning, 36, 38, 44, 60, 111
psychiatrists, 118
psychiatry, 139
psychological, 107, 128, 131
psychological value, 107, 128
psychologists, 118
psychology, vii, 137
puberty, 36, 111
punishment, 99, 113, 117, 123, 125
pyramidal, 105

Q

quality improvement, 121
questioning, 92, 116

R

radial glia, 40, 41, 68
random, 103
random access, 103
range, 37, 38, 44, 46, 53, 58, 60, 71, 74, 88, 89, 91
raphe, 23
ras, 81
rat, 6, 23, 29, 30, 75, 76, 124, 125
ratings, 119
rational expectations, 86
rationality, 86, 91, 119, 129
rats, 62, 63, 66, 73
reality, 98, 103, 113, 118, 123, 124, 125, 127, 141
reasoning, 90, 110, 114, 122, 129, 137
receptive field, 48
receptors, 7, 16, 26, 29, 47, 62, 68, 71, 87
recognition, 22, 30, 51, 95, 117, 121
recollection, 123

reconciliation, 106
recovery, 49, 66, 75, 76, 81
recreational, 39
recycling, 6, 8, 14, 15, 19
redistribution, 81
reduction, 23, 27, 37, 42, 44, 80
redundancy, 60
regional, 35, 37, 38, 41, 42, 70, 83
Registry, 67
regression, 39
regulation, vii, 2, 25, 29, 30, 40, 72, 78, 104, 107, 138
rehabilitate, 48
rehabilitation, 48, 64, 72, 76
reinforcement, 38, 87, 99, 108, 135
reinforcement learning, 135
rejection, 88
relationship, 40, 57, 78, 80, 82, 91, 95, 103, 109, 117, 120
relationships, 40, 80, 87, 88, 95, 99, 106, 111, 114, 115, 117, 124, 125
relative size, 37
relatives, 88
relativity, 131, 137
relevance, 4, 7, 104
reliability, 67, 93, 101, 116
Reliability, 64
religion, 39
religious, 58
remodeling, 45, 66, 115
repair, 113
replication, 88
reproduction, 99, 100
research, vii, viii, 33, 46, 48, 51, 54, 61, 80, 85, 90, 115, 126, 128, 129, 135
research design, 90
resentment, 124
reserve capacity, 48, 75
residues, 5, 7
resilience, 100
resistance, 35, 39, 60, 100
resolution, 103
resources, viii, 33, 100, 108, 115, 126, 129
restoration, 26
reward pathways, 87

rewards, 106, 113, 115
rice, 50
rigidity, 99, 100, 117, 118, 119, 125, 128
risk, 11, 39, 50, 55, 56, 57, 60, 62, 63, 66, 67, 68, 77, 79, 81, 93, 94, 103, 107, 129, 139, 141
RNA, 7, 16
RNAi, 8, 10
rodent, 29, 67
Royal Society, 138

S

safety, 124
sample, 82
scaffold, 40
Schmid, 35, 75
science, vii, viii, 33, 90, 91, 101, 128, 136, 139
scientific, vii, 35, 52, 91, 92, 94, 96, 99, 107, 114, 116, 119, 125, 127, 128
scientific knowledge, 91, 125
scientific method, 127
scientific understanding, 91
scores, 58
searching, 105
secretion, 8, 10, 16, 22
secrets, 19
segmentation, 83
selecting, 115
Self, 124, 135
self-awareness, 140
self-esteem, 123, 124, 126, 133
semantic, 78, 91, 94, 106, 108, 109, 116, 140, 141
semantic memory, 140, 141
semantics, 89, 90, 91, 94, 96, 101, 106, 107, 110, 116, 125, 128, 135
senile, 11
senile plaques, 11
sensitivity, 30, 57
sensory systems, 37
sequencing, 103
series, 65, 104, 119
serine, 8

serotonergic, 26
serotonin, viii, 21, 22, 23, 26, 28, 30, 133
Serotonin, 28
serum, 59
services, iv
sex, 71
shape, 46
shelter, 98
shock, 124
signal transduction, 17, 25
signaling, 28, 41, 45, 46, 47, 48, 77, 139
signaling pathway, 46
signaling pathways, 46
signalling, 26, 72, 81
signals, 6, 47
siRNA, 8
sites, 6, 29
skills, 93, 95, 106, 111, 114, 115, 116, 121, 129
skin, 48
Skinner, B. F., 140
smoothing, 36
social, 39, 65, 87, 88, 89, 90, 95, 98, 99, 100, 101, 104, 106, 107, 111, 116, 130, 131, 132, 135, 137, 138, 139
social attributes, 87
social behavior, 87, 88, 132, 138
social cognition, 130, 137, 139
social development, 100
social group, 88, 89, 99
social perception, 106
social psychology, 138
social rules, 100
social structure, 101
social systems, 99
social work, 104
socialization, 86, 87, 95
socially, 129
society, 128, 129
socioeconomic, 35
socioeconomic status, 35
SOD1, 7
sodium, 4, 19
software, 102, 110, 114, 115, 122
solutions, 114, 116

somatosensory, 25, 28, 65, 107
sorting, 6, 7, 11, 12, 16
sounds, 118
spatial, 18, 93, 104
spatial location, 93
specialization, 137
species, 86, 87, 88, 95, 99, 121, 136
specificity, 57, 97, 98, 127, 132
SPECT, 71
spectra, 108, 120
spectrum, 91, 96, 100, 124
speech, 93, 94, 105, 107, 112
spine, 47, 66, 71
spines, 66
sporadic, 11, 19
sprouting, 35
SR, 28, 29, 30
stability, 42
stages, 25, 37, 44, 53, 74, 100, 111, 112
standards, 101, 119, 124
statistical analysis, 74
stereotyping, 117
stimulus, 86, 90, 93, 103, 105, 106, 118
stomach, 22
storage, 22, 23, 28, 61, 102, 109, 111, 112, 121, 126
strategies, vii, 2, 95, 99, 103, 108, 125
strength, 41
stress, 88, 115
striatum, 25, 103, 132
stroke, 73, 81
strokes, 65
structural changes, 64
structural protein, 45
Subcellular, v, 1, 5, 6
subcortical structures, 42, 82
subgroups, 89
subjective, viii, 85, 90, 91, 92, 93, 94, 96, 97, 98, 99, 101, 103, 105, 107, 108, 109, 111, 112, 115, 116, 118, 120, 123, 124, 126, 128
subjectivity, 110
substance abuse, 118
substantia nigra, 27
substrates, 4, 19, 23, 137

subsymbolic, vii
subventricular zone, 25
successful aging, 48, 60
Sun, iii, 16, 28, 69
superoxide, vii, 2, 7, 12, 19
superoxide dismutase, vii, 2, 7, 12, 19
superstitious, 92, 114
survival, 28, 66, 95, 99, 100, 126
survivors, 58
switching, 105
symbolic, vii, 90, 125
symbols, vii, 95
symptom, 35, 39, 42
symptoms, viii, 33, 34, 35, 39, 40, 44, 51, 53, 54, 55, 57, 58, 59, 60, 61, 88
synapse, 64, 68, 73
synapses, viii, 21, 25, 38, 45, 49, 52, 63
synaptic plasticity, 45, 46, 61, 65, 67, 73, 77, 81, 82
synaptic transmission, 22
synaptic vesicles, viii, 21, 22, 23, 26, 27, 29
synaptogenesis, 35, 36, 40
synaptophysin, 73
synchronous, 37
syndrome, 3, 49
syntax, 93
synthesis, 24
synthetic, 23
systematic, 105
systems, 89, 91, 98, 100, 104, 105, 109, 111, 112, 117, 122, 128, 133, 135

T

tactics, 116
tangles, vii, 1
targets, 11, 25, 69
task demands, 106, 107
tau, vii, 1, 13, 57, 66, 73
teachers, 123
teaching, 129
technological, 96
technological advancement, 96

Index   161

temporal, 42, 54, 55, 56, 57, 62, 65, 66, 69, 70, 103, 104, 105, 108
temporal lobe, 42, 56, 66, 70, 103, 104
terminals, 6, 66
thalamus, 23, 105
theoretical, 34, 52
theory, 91, 95, 100, 116, 126, 131, 134, 137, 139, 141
therapeutic, vii, 2, 11
therapeutics, 16
therapy, 17, 27, 68, 72, 81, 131, 141
thinking, 90, 91, 92, 100, 110, 112, 113, 114, 115, 116, 117, 121, 122, 127, 130, 133, 135, 136, 137, 138
threatening, 105
threats, 115
threshold, 34, 39, 60
time, 35, 51, 58, 94, 95, 96, 97, 100, 102, 103, 109, 114, 115, 116, 120, 127, 128, 130
time warp, 114
timing, 111
tissue, 7, 35
toddlers, 62
tonic, 137
top-down, 104, 107, 133
toxic, viii, 21, 22, 23, 27
toxic effect, 27
toxicity, 23, 27, 29, 30
toxin, 22, 27, 39
tracking, 126
trade, 132
traffic, 16
training, 72, 113
traits, 40, 79
trajectory, 60
trans, 18
transcranial magnetic stimulation, 48, 64
transcription, 46, 47, 72
transcription factor, 72
transcription factors, 72
transcriptional, 7
transfer, 98, 125
transformation, 129
transgenic, 4, 8, 16, 64, 71

transgenic mice, 4, 8, 16, 64
transition, 54, 61
translation, 46
translational, 12
translocation, 5, 22
transmembrane, vii, 1, 3, 7, 12, 18
transmission, 94
transparent, 116
transport, vii, 2, 5, 6, 7, 8, 9, 10, 11, 14, 15, 18, 28, 127
traumatic brain injury, 63, 75
trend, 36
triggers, 88
trust, 117
turnover, 45
twins, 62, 76
tyrosine, 24
tyrosine hydroxylase, 24

U

ubiquitin, 6, 16
ultrasound, 35
uncertainty, 92, 113, 124, 126, 127, 129, 140, 141
underlying mechanisms, 60
universal grammar, 93
universe, 117
updating, 103

V

valence, 88, 104, 113
validation, 51
validity, 67, 93, 114
values, 87, 89, 91, 106, 117, 124, 128
variability, 49, 125
variable, 38, 46, 60, 98, 122
variables, 54, 87, 95, 117, 119, 122
variance, 58, 96, 113
variation, 41, 59, 60, 62, 87
vascular, viii, 33, 34, 50, 78, 83
vascular dementia, 50, 83
vascular risk factors, 78

vasopressin, 87, 131, 132, 133
ventricular, 25, 35, 40, 41, 72, 75
ventricular zone, 40
verbal fluency, 57
vesicle, 9, 22, 28, 30
violence, 88, 131
vision, 93
visual, 80, 81, 93, 94
visual processing, 94
vocabulary, 97
voles, 132
voxel-based morphometry, 64
vulnerability, 15, 29, 44, 53, 76
Vygotsky, 86, 89, 94, 141

W

walking, 24
Washington, 64, 130, 134, 136, 139, 140, 141
water, 39, 98
Watt Governor, vii
wealth, 105
white matter, 36, 77, 78, 82
wisdom, 125
Wnt signaling, 41

women, 29, 79
word meanings, 97
working memory, viii, 57, 83, 85, 90, 102, 103, 104, 108, 109, 115, 132
worldview, 117
writing, 94, 125

X

X-linked, 81

Y

yield, 110, 115, 128
Y-maze, 82
young adults, 48

Z

zinc, 7
Zn, 19
zoology, 128